Environmental Accounting and Reporting

Theory, Law and Empirical Evidence

Environmental Accounting and Reporting

Theory, Law and Empirical Evidence

SHUCHI PAHUJA
Associate Professor
PGDAV College, University of Delhi, Delhi

NEW NC CENTURY

New Century Publications
New Delhi, India

NEW CENTURY PUBLICATIONS
4800/24, Bharat Ram Road,
Ansari Road, Daryaganj,
New Delhi – 110 002 (India)

Tel.: 011-2324 7798, 4358 7398, 6539 6605
Fax: 011-4101 7798
E-mail: indiatax@vsnl.com • info@newcenturypublications.com
www.newcenturypublications.com

Editorial office:
LG–7, Aakarshan Bhawan,
4754-57/23, Ansari Road, Daryaganj,
New Delhi – 110 002

Tel.: 011-4356 0919

First Published: **September 2009**

ISBN: **978-81-7708-220-3**

Published by New Century Publications and printed at Salasar Imaging Systems, New Delhi.

Designs: Patch Creative Unit, New Delhi.

PRINTED IN INDIA

Dedicated to my parents

Shri Surender Raj Mahajan
and
Shrimati Kanta Mahajan

For instilling and fostering in me the values of hard work and perseverance

About the Book

The last two decades have witnessed an increasing global concern for the harmful long-term impact of industrial activities on the environment and subsequently on financial performance of a company and economic growth of a country. This has led to a growing demand from various stakeholders for measurement of a company's environmental activities and impacts, and subsequent public disclosure of this information. As a result, a new area of accounting has emerged, viz. environmental accounting.

Environmental accounting at corporate level deals with identification, measurement, recognition and disclosure of environmental costs, liabilities and contingencies in the financial records of a company for the benefit of various internal and external stakeholders.

This book provides a broad introduction to the whole area of environmental accounting and reporting (EAR). It covers environmental accounting, environmental reporting, and environmental auditing together as these are interrelated and form a very important part of a company's comprehensive environmental management system. An attempt has also been made to assess the need for a separate regulatory framework to address these issues more specifically.

Author's Profile

Dr. Shuchi Pahuja is Associate Professor at PGDAV College, University of Delhi, Delhi. She obtained her doctoral degree from the Department of Commerce, Delhi School of Economics, University of Delhi in 2006. She has teaching experience of about one and a half decades.

Dr. Pahuja has presented papers in various national and international conferences including 29th Annual Congress of European Accounting Association (EAA), held in Dublin, Ireland in March 2006. She is a member of Social Responsibility Research Network (SRRN) of UK.

Preface

Accounting and reporting for the environment has become an important dimension of corporate external information system in 21st century. The increasing pressure on companies to be responsible to the society has influenced them to operate in an environmentally responsible manner. 'The impact of environmental performance of an enterprise on its financial health' has become a matter of growing concern to investors, creditors, government and public at large. As various stakeholders demand greater disclosure of environmental impacts and performance, a large number of companies all over the world, have started reporting on these issues. In many countries (like Denmark, the Netherlands, Spain and Australia), disclosure of some environmental information (in or outside the financial statements) has also been made mandatory. But no accounting body has yet issued any specific guideline or standard on the issue. Various research findings have shown that in the absence of any accounting standard or guideline on the issue, the environmental disclosures made by companies generally lack quality and consistency in reporting. In fact, there is a need to improve the conceptual underpinning and thus, the quality of environmental accounting and reporting. It requires development of a conceptual framework of environmental accounting and reporting, and subsequently a standard based on it, containing objectives, general assumptions, qualitative characteristics and guidelines for companies.

Disclosure of social and environmental information has also been an area of substantial research in the past few decades. Since 1990, a number of studies have been made to examine the environmental reporting practices of the companies. Though some international surveys (e.g. KPMG 2002) examined environmental information disclosures by concerns from different countries, most of the country-specific works focused on the green reporting practices of corporations mainly from Europe and USA. However, empirical evidences from the developing countries are very small in number. Most of the existing studies analysed communication of environmental information by companies in general, without concentrating on quality of reporting. Very few studies have examined the impact of various corporate- and industry-related variables on ecological disclosures. Environmental accounting and auditing areas are yet to be explored (particularly in India). A preponderance of existing works are based on secondary data and lack insight into various conceptual and theoretical considerations involved in accounting and reporting on environment. The present work has attempted to bridge these gaps.

The book explains various theoretical and conceptual considerations involved in environmental accounting and reporting (EAR). It provides an insight into regulatory framework relating to EAR in various countries. The environmental accounting and reporting practices followed by the corporate sector in India have also been examined in the book.

Some salient features of the book are as follows:

It explains theoretical concepts of EAR by referencing them to a conceptual framework of corporate accounting and reporting.

The book analyses in detail EAR laws in various countries and highlights the need for a specific regulatory framework on EAR (including accounting guidelines, principles and standards) in India.

Various empirical studies conducted all over the world including India have been thoroughly reviewed in the book. The book provides comprehensive literature on the issue which would be useful to researchers, professionals and academicians in the area.

An attempt has been made in the book to find evidence on existence of corporate environmental accounting and reporting expectation gap in India. The book examines EAR practices of large manufacturing companies in India and their environmental information audit systems.

To ascertain the extent of mandatory and voluntary disclosure of environmental information by large manufacturing companies in India, their annual reports have been examined. In addition, to explain why the levels of disclosures vary across the companies and industries, a multivariate analysis has been done.

It is hoped that observations and suggestions made in the book would be beneficial for the investors, creditors, government and other stakeholder. These will also help companies and business executives including industry accountants and auditors in dealing with environmental matters in the books of accounts. Findings of the book will be useful to the research scholars, research institutions, professional accounting bodies and business schools.

The book is divided into 7 Chapters.

Chapter 1 is introductory in nature and provides a context to the subject by discussing in brief the key concepts, the objectives, and scope and methodology used in the study.

Chapter 2 presents theoretical considerations relating to environmental accounting, reporting and auditing. It covers definitions, benefits, and conceptual framework of EAR.

Chapter 3 examines existing regulatory frameworks of EAR at national and international level and the problem areas.

In Chapter 4 some of the existing studies on EAR conducted in India and abroad have been reviewed. This chapter also gives implications and relevance of past works and the rationale for the present work on EAR in India.

Chapter 5 examines evidence on existence of corporate environmental accounting and reporting expectation gap in India with the help of primary data collected from the chartered accountants and the top executives working in the large manufacturing companies in India.

Environmental disclosure practices followed by the Indian corporate sector have been analysed by examining annual reports of the selected companies in Chapter 6. The Chapter also gives results of multivariate analysis, showing relationship between environmental information disclosure and selected industry and company-specific variables.

Finally Chapter 7 presents summary and conclusions and makes recommendations for the benefit of professional accounting bodies, regulatory authorities, managers, and other stakeholders.

I wish to place on record my profound gratitude to Professor S.C. Bansal for his keen interest, critical comments and valuable suggestions to improve the present work. Thanks are also due to Dr. H.L. Bhatia (formerly Principal) and Dr. Ashok Gupta both of PGDAV College, University of Delhi, Delhi for constantly motivating me to complete this work.

I highly appreciate the affection and moral support received from my family members, in particular my father Shri S.R. Mahajan, mother Shrimati Kanta Mahajan, husband Shri Kuldip Pahuja and children Baby Ananya and Baby Archisha.

Finally, I thank Mr. Sandeep Sury, Chief Executive, New Century Publications, New Delhi and his anonymous editors for improving the format and presentation of the material contained in this work.

Delhi **Shuchi Pahuja**
August, 2009

Acronyms and Abbreviations Used in the Text

AAA	American Accounting Association
ACBE	Advisory Committee on Business and Environment (UK)
ACCA	Association of Chartered Certified Accountants (UK)
AICPA	American Institute of Certified Public Accountants
APB	Accounting Principle Board
AS	Accounting Standards (India)
BATNEEC	Best Available Techniques not Entailing Excessive Cost
BiE	Business-in-the-Environment
BSE	Bombay Stock Exchange
CA	Chartered Accountants
CAS	Corporate Accounting Standards
CER	Corporate Environmental Report
CICA	Canadian Institute of Chartered Accountants
CS	Company Secretary
CSC	Canadian Standards Committee
CWA	Cost and Works Accountants
D/E	Debt-equity Ratio
D_{BH}	Dummy for Business House
D_{FA}	Dummy for Foreign Association
DGM	Deputy General Manager
DII	Department of Institutional Investors
D_{MP}	Dummy for Most Polluting Industry
DoE	Department of Environment
D_S	Dummy for Sector
DSE	Delhi Stock Exchange
ESR	Exports to Sales Ratio
EAR	Environmental Accounting and Reporting
EDS	Environmental Disclosure Score

EITF	FASB's Emerging Issues Task Force
EMA	Environmental Management Accounting
EMAS	Eco-Management and Audit Scheme of EU
EMS	Environmental Management System
EP	Environmental Performance
EPA	Environmental Protection Agency (USA)
EU	European Union
FASB	Financial Accounting Standards Board (USA)
FEE	Federation des Experts Comptables Europeens (Europeans Federation of Accountants)
FMCG	Fast Moving Consumer Goods
FRS	Financial Reporting Standard (UK)
GAAP	Generally Accepted Accounting Principles
GEMI	Global Environmental Management Initiative
GHG	Green House Gases
GM	General Manager
GRI	Global Reporting Initiative
IAPC	International Auditing Practices Committee
IASC	International Accounting Standards Committee
ICAI	Institute of Chartered Accountants of India
ICC	International Chamber of Commerce
ICWA	Institute of Cost and Work Accountants (India)
IIIEE	International Institute for Industrial Environmental Economics (Lund University)
IPC	Integrated Pollution Control
IPPC	Integrated Pollution Prevention and Control Directive (EU)
IRRC	Investment Responsibility Research Centre, USA
ISAR	United Nations' Intergovernmental Working Group of Experts on International Standards of Accounting and Reporting
ISO 14000	International Specification and Guidelines for Using Environmental Management Systems (since 1996)

KSEC	Korean Securities Exchange Commission
MoEF	Ministry of Environment and Forests
MNCs	Multinational Corporations
N.A.	Not Answered
NGOs	Non-Governmental Organizations
NPI	National Pollutant Inventory (NPI)
NRTEE	National Round Table on the Environment and the Economy
OLS	Ordinary Least Squares
P	Profitability
P&L A/c	Profit and Loss Account
PCB	Pollution Control Board
S	Size
SEBI	Securities and Exchange Board of India
SEC	Securities and Exchange Commission (USA)
SFAS	Statement of Financial Accounting Standards (US)
SPCB	State Pollution Control Board
SRI	Sustainability Reporting Initiatives
TRI	Toxic Release Inventory (USA)
UN	United Nations
UNEP	United Nations Environment Programme
WRI	World Resource Institute

Contents

1

Environmental Accounting and Reporting: An Introduction

Accounting is the process of identifying, measuring, recognising and communicating financial information to the diversified stakeholders. Over the past few decades, there has been a shift in emphasis regarding the main purpose of accounting. The modern accounting is not merely concerned with record keeping and reporting of information to the investors, but it aims at fulfilling information needs of a wide range of internal and external stakeholders. It is now considered as a service activity. "Its function is to provide quantitative information primarily financial in nature about economic activities that is intended to be useful in making economic decisions." [1] The basic objective of modern accounting is, therefore, to provide useful information about the activities of an entity to present and potential investors, creditors and other stakeholders for their use in making relevant investment, credit and other similar decisions.

Since late eighties, due to growing public concern about the alarming impact of industrial activities on nature (e.g., disastrous consequences of atmospheric pollution, oil spills, chemical and other toxic discharges and depletion of scarce natural resources), companies are under increasing pressure from government and society to reduce adverse environmental impacts of their activities by making efficient use of scarce resources and by using cleaner production technologies. Rapidly changing laws and increasing attention given to environmental issues by the regulatory authorities all over the world make it necessary for a company to examine impact of its activities on the environment. The performance of an organisation is now being judged not only on the basis of its financial results, but also with regard to its contribution to protect and improve the environment. There is an increasing emphasis that environmental issues have considerable implications for financial and investment decisions. They have become an important variable in the models used by investors and creditors to determine the risk associated with their investment. Before making any major financial transaction (such as investment, capital budgeting, or corporate restructuring), it is necessary to take into account a company's environmental performance and its impact on the concern's trading results and financial position. In the evolving scenario, the need for accounting and reporting on the environment has largely been felt. As a result, accounting of environmental issues and their disclosure in the annual reports or by other mediums has become an important part of corporate accounting and reporting systems. This book aims at presenting various theoretical and legal considerations involved in environmental accounting and reporting (EAR) together with empirical evidences from different countries.

The present Chapter is introductory in nature. It aims at providing the rationale for the present book. It is divided into three Sections. Section 1.1 covers various concepts relating to EAR used in the book. In Section 1.2, need for environmentally sensitive accounting in India has been analyzed in view of changing national and international scenario. Finally, Section 1.3 outlines the objectives of the present book.

1.1 Various Concepts Relating to EAR

As mentioned above, there is an emphasis on environmentally sustainable industrial development these days. Environment refers to the aggregate of all the external conditions and influences affecting the life and development of living beings. According to Environment (Protection) Act, 1986, "Environment includes water, air and land and the inter-relationship which exists among and between water, air and land, human beings, other living creatures, plants, micro organism and property." [2]

To measure environmental activities of an organisation and their resulting impact on business and the society, it is necessary to account for the concern's environmental costs, benefits, assets and liabilities. "Environmental accounting refers to the estimation of and public reporting of environmental liabilities and financially material environmental costs." [3] Broadly speaking, it involves preparation, verification and disclosure of information relating to environmental activities and performance of an organisation.

Due to growing social and legal pressures and increasing judicial intervention, there has been a growing demand for disclosure of environmental policies, practices and performance of a company to the interested stakeholders in or outside the concern. "Environmental reporting is the term commonly used to describe the disclosure by an entity of environmentally related data, verified (audited) or not, regarding environmental risks, environmental impacts, policies, strategies, targets, costs, liabilities or environmental performance to those who have interest in such information as an aid to enabling/enriching their relationship with the reporting entity, via either the annual report and accounts package; a stand-alone corporate environmental report(CER); a site-cantered environmental statement, or some other medium (e.g. staff newsletter, video, CD Rom, internet site)." [4]

Reporting of environmental information to the external stakeholders gives rise to the need for verifying these statements, i.e. environmental audit. Broadly environmental audit is verification of environmental management system and the related policies and practices followed by a concern. "Environmental audit can be defined as a management tool comprising systematic, documented and periodic evaluation of how well environmental organization management and equipment are performing with an aim of helping to regularize the environment." [5] Thus, environmental audit is a broad term covering a number of activities. But for the purpose of this study, it has been taken only as an audit of environmental information supplied by a company in the annual report or environmental statement. Environmental audit has been defined as a systematic process of objectively obtaining and evaluating the evidence relating to the performance of an organisation as reflected in the environmental reports.

1.2 Recent Developments

This Section briefly examines the need for environmentally sensitive accounting in India in view of changing national and international scenario. [6] It explores how the current role of accounting in India must be extended and adapted to meet the changing requirements of various stakeholders.

Growing legal and social concerns about mounting natural calamities and other environmental problems have given rise to the need for incorporating environmental issues in the financial statements of a company. As a result, a new area of accounting has emerged known as environmental accounting. The United Nations Inter Governmental Working Group of Experts on International Standards of Accounting and Reporting (ISAR) began to focus on the subject of EAR in the late 1980s. The growth in interest and activity in environmental accounting in the past two decades is astonishing. From the most marginal and irrelevant subset of social accounting at the beginning of 1990s, it has expanded to become something which is now seen as an essential element in any organization's environmental responsibility. Starting from 1989, ISAR has issued number of recommendations on EAR. Apart from ISAR, different types of supra-national bodies (e.g. FEE, European Union) have also shown interest in environmental reporting and some of these have issued guidelines in this connection. Professional accounting bodies all over the world have this item on their agenda. But till date, legal authorities, standard setting bodies and other regulators have not come to a consensus regarding the conceptual framework of environmental accounting and its disclosure.

Significant developments in Ind an economy that had implications for EAR began in early 1990s. Following the opening up of Indian economy in 1990-91, Indian companies have steadily increased their level of activity in the global arena. The number of Indian companies investing abroad has been growing consistently. The international concern regarding the promotion of trade in environmental goods and services required introduction of the concepts of clean technologies, sustainable vision and sustainable business. These sustainability initiatives require calculation of environmental indicators for appraisal of various activities and projects and their subsequent disclosure in different reports including annual report of the company for the benefit of various stakeholders. Indian companies are also facing strong international competition. These international companies are disclosing non-financial information including environmental information leading to an enhanced expectation from Indian companies to act responsibly towards the environment and be accountable to the society beyond the traditional role of providing financial account to the shareholders.

Regulation of environmental issues is growing rapidly in all countries of the world. The Government of India is also promoting more and more regulations to protect the environment and the community in general. Violation of these statutes can result in significant fines, remediation costs and even closure of business. Because of more stringent environmental laws, in all the major business decisions (like financing, capital budgeting, working capital management, cost controls, project planning and control, mergers or corporate restructuring), environment has become one of the critical factor. With this legislative trend, assessment of environmental performance and environmental impacts of a business has become important to managers not only for reasons of social concern but because they represent very real liabilities faced by corporations.

Over the past two decades, there has also been constant growth of concern about environmental issues among investors (particularly institutional investors), creditors, customers, NGOs, and society at large. The measurement and reporting of environmental performance is becoming increasingly important to satisfy the growing expectations of various stakeholder groups. With growing legal and social awareness, there has been a significant increase in environmental costs and liabilities, to the extent that they have become a significant part of total costs and liabilities. Concealing them or their improper disclosure might expose these interested stakeholders to unknown levels of risk.

Public concerns the world over have resulted in building mechanisms such as the Intergovernmental Panel on Climate Change (IPCC), the Kyoto Protocol for Greenhouse gas (GHG) reduction and environmental taxes. Many countries are now considering to control green house effects and CO_2 targets through regulation of business entities using carbon emission rationing system. Many related accounting issues have arisen due to carbon trading which must be answered by accountants in India. Moreover accountants are required to pay attention to the taxation aspects of environmental issues. Recent discussions on 'triple-bottom-line reporting' or 'Sustainability Reporting Initiatives (SRI)' in the financial markets have also underlined a need for adoption of environmentally sensitive accounting by corporate sector in India.

The above discussion shows that changes in national and international environment have posed many new challenges for accountants in India. However, the accounting profession in India has yet to display significant role in sustainability. Environmental accounting is in preliminary stage in India. Even when businesses become more aware of environmental expenses, benefits and liabilities and the growing demand for environmental information in the annual accounts, they are often uncertain about how to account for them. Neither the company law nor the accounting standards/guidelines issued by the ICAI prescribe accounting and disclosure norms for the environment related issues in the corporate financial statements or environmental reports. Various research findings have shown that in the absence of accounting standards or guidelines on the issue, the disclosure of environmental information by companies in India generally lacks quality and consistency. There is an urgent need, therefore, to adopt a more disciplined and systematic approach towards accounting and disclosure of environmental information so as to provide useful, objective and timely environmental information to the concerned stakeholders. In view of current developments in the financial market place, accounting bodies in India should take a broader view of financial standards and require an explicit consideration of social and environmental issues in the financial statements. In fact, a complete greening of accountancy is required in India which involves a reappraisal of how to identify and measure the relevant environmental costs and liabilities and to analyze the extent to which environmental activities and performance can influence managerial decisions and financial results of a company.

Objectives of the Study

In the light of observations made in the previous Section, it was felt that there is a need to have insight into various issues relating to EAR. The necessity for comprehensive EAR laws in India and the role of accounting in improving environmental performance of an organization is also required to be understood. The present book is an attempt to bridge some of the research gaps in the area in India. It identifies various theoretical and legal considerations involved in EAR. On the basis of empirical evidences from India and abroad, the study also makes some recommendations to improve the quality of external corporate environmental reporting in India.

More specifically, the book has attempted to achieve the following objectives:

- To examine various conceptual and theoretical considerations involved in EAR.
- To analyse in detail EAR laws in various countries and to highlight the need for a specific regulatory framework on EAR (including accounting guidelines, principles and standards) in India.
- To review various empirical studies conducted all over the world (including India) and to find the need for a comprehensive global study in India on EAR.
- To examine EAR practices of large manufacturing companies in India and their environmental information audit systems. An attempt has also been made to find evidence on existence of corporate EAR expectation gap in India.
- To ascertain the extent of mandatory and voluntary disclosure of environmental information by large manufacturing companies in India in their annual reports.
- To explain why the levels of disclosures vary across the companies and industries with the help of multivariate analysis.

Based on existing available literature (theory and findings of past research works) on EAR, various propositions relevant to the above objectives have been made and tested in the present book. The propositions made, research methodology used to test them, and the findings of various tests have been discussed at relevant places in subsequent chapters. It is hoped that observations and suggestions made in the study would help the professional accounting bodies and regulatory authorities to develop some standard or guidelines on the issue. It would also help the companies and management in handling 'green matters' effectively and efficiently in the financial statements.

Notes

1. Accounting Principles Board (APB), Statement No.4, *Basic Concepts and Accounting Principles Underlying Financial Statements of Business Enterprise*, AICPA, (1970), Para 40.
2. Environment (Protection) Act, 1986 of India.
3. Environmental Protection Agency (EPA), '*An Introduction to Environmental Accounting as a Business Management Tool: Key Concept sand Terms*', US: EPA primer, Washington DC: EPA, (1995a), p.28.
4. UN-ICAI Seminar Report, '*Environmental Accounting and Reporting*', conducted at New Delhi, (August 2-4, 1999), p. 6.2.
5. United Nations Environment Programme (UNEP), '*Environmental Auditing*', Technical Report Series No. 2, UNEP, Industry and Environment office, (1990).
6. For detailed analysis on 'Need for EAR Standards in India', refer to Section 3.2.

2

Environmental Accounting and Reporting: Concepts and Theoretical Framework

As mentioned in the previous Chapter, these days there is an emphasis on environmentally sustainable industrial development. It has increasingly been realised that economic development without environmental considerations can cause serious environmental damage, in turn, impairing the quality of life of present and future generations. It is, therefore, necessary that timely steps are taken by business and industry to control pollution and to conserve and protect environment. Research has shown that eco-efficiency is not only beneficial for the society, but can also lead to resource savings, cut in production costs and increase in profits for business. To measure environmental activities of an organisation and the resulting impact on business and the society, it is necessary to account for the company's environmental costs, benefits, assets and liabilities. There is also a need to disclose to the community at large, data concerning a company's environmental policies, environmental management programmes and the impact of environmental performance on its financial results.

As a result, accounting and disclosure of environmental matters has been rapidly emerging as an important dimension of corporate accounting and reporting practices. Since mid 1990s, some professional bodies in USA, UK, Belgium, Denmark and Australia have started taking some initiatives in this regard, but they are still in the process of examining the need for a separate conceptual framework or guidelines on environmental accounting and reporting (EAR). No country has yet formulated an accounting standard on this issue.

In this Chapter, an attempt has been made to discuss some theoretical considerations relating to environmental accounting, reporting and auditing. Before presenting these aspects, it is necessary to give a general idea about the term 'environment' and, 'environmental management system'. Therefore, Section 2.1 explains the concepts of 'environment' and 'environmental management system (EMS)'. Section 2.2 discusses definition of environmental accounting, its benefits and other specific issues relating to accounting for environment. Section 2.3 elaborates various aspects of environmental reporting. Section 2.4 briefly puts up the concept of auditing environmental information supplied by a company in annual reports or environmental statements. Finally, Section 2.5 summarises the Chapter.

2.1 Environment and Environmental Management System

In this Section, an attempt has been made to explain the concepts of 'environment' and 'environmental management system (EMS)'. While sub-section 1 presents some important definitions of the term environment, sub-section 2 covers the concept of EMS.

2.1.1 Environment: The term environment refers to all the conditions that surround us. The term has been defined differently by various authors. Some of these definitions are:

"The physical surroundings, conditions, circumstances etc. in which a person lives... the totality of the physical conditions on the earth or part of it." [1] "The surroundings in which an organism lives, which are affected by physical and chemical factors and by other organisms." [2]

These definitions are very broad and define environment in a general way. In the context of business organisations, environment can be defined as all the conditions that surround a business concern. These conditions include external as well as internal environment of the organisation. The external environment consists of natural physical environment (like land, raw materials and energy) and general environment (like economic, social, cultural, political and legal environment). Internal environment consists of all living and non-living things as well as environmental forces working within a concern (e.g., plant and machinery, building, human beings and organisation culture).

According to Environment (Protection) Act, 1986, "Environment includes water, air and land and the inter-relationship which exists among and between water, air and land, human beings, other living creatures, plants, micro organism and property." [3] This definition covers natural physical environment only. The present book also mainly focuses on natural physical environment and, therefore, ignores non-physical, general, industrial and internal environment. Environment has been taken as natural physical surroundings and includes air, water, land, flora, fauna and non- renewable resources such as fossil fuels and minerals.

2.1.2 Environmental Management System (EMS): In recent years, the environmental crisis has become a global issue. The concept of sustainable development is gradually being recognised by all concerned. Business and industries are increasingly realising their role in sustainable development and ultimately a new area of management is fast emerging known as 'Corporate Environmental Management System'. Any EMS may be considered as a part of the overall management system. It represents the organisational structure, responsibilities, processes and preconditions for the implementation of a company's environmental policy. ISO–14001 specifically governs EMS; the structure, policies and procedures by which corporations manage their environmental impacts. It is a methodology for environmental management. It provides necessary input for any organisation to develop and implement a cost-effective system of management by introducing environmental audit.

According to ISO-14001, "EMS is that part of the overall management system which includes organizational structure, planning activities, responsibilities, practices, procedures, processes and resources for developing, implementing, achieving, reviewing and maintaining the environmental policy." [4] Thus, EMS is a framework or method of guiding an organisation to achieve and sustain performance in accordance with established environmental policy and goals and also keeping in mind changing regulations, social and environmental pressures. An efficient system of environmental management helps management in complying with environmental laws, acknowledging environmental risks as well as controlling them and reducing adverse impact of its activities on environment.

2.2 Environmental Accounting

Environmental accounting is an emerging concept of accounting. During the last two decades, the accounting profession worldwide has made various deliberations on this topic. Because of these deliberations, accounting for the environment is now seen as an essential element of an organisation's environmental management system. In this Section, an attempt has been made to present theoretical considerations relating to environmental accounting. This Section is further divided into 3 sub-sections. Sub-section 1 deals with the definition of environmental accounting. Sub-section 2 lists benefits of adopting environmental accounting system. Sub-section 3 discusses some specific issues relating to accounting of environmental information.

2.2.1 Definition of Environmental Accounting: Various professional bodies and authors have defined environmental accounting differently. At macro level, environmental accounting can be defined as, "one of the methods that takes into account environmental resources and services and changes therein and measures their effects on GNP and NNP to reveal true maximum income (True Net Capital Formation) which a nation can consume, while maintaining a sustainable development and growth, without jeopardizing the interests of the present and future generations as well as our neighbours." [5]

At corporate level, environmental accounting is sometimes taken as a subset of social accounting. Social and environmental accounting is "the process of communicating the social and environmental effects of an organisation's economic actions to particular interest groups within society and to society at large. As such, it involves extending the accountability of an organisation (particularly a company's) beyond the traditional role of providing a financial account to the owners of capital, in particular, shareholders." [6] Such an extension of the responsibilities of an organisation is based upon the assumption that companies do have wider responsibilities than simply to make money for shareholders.

Some authors have defined environmental accounting at management accounting level also. Environmental Management Accounting (EMA) is concerned with the presentation of data about environmental costs, environmental liabilities and performance to management, so that it can also be considered while making decisions. EMA leads to improved decision-making by identifying hidden or misallocated internal and external environmental costs and allocating those to particular products, processes or facilities.

Thus, the term environmental accounting has many meanings and uses. It can support national income accounting, financial accounting or internal business management accounting. Environmental Protection Agency (EPA) of USA has explained environmental accounting at these 3 levels as under: [7]

- **Environmental accounting in the context of national income accounting** refers to natural resource accounting, which can entail statistics about a nation or region's consumption, extent, quality and value of natural resources both renewable and non-renewable.
- **Environmental accounting in the context of financial accounting** usually refers to the preparation of the financial environmental reports for external audiences

using GAAP. It includes estimation and public reporting of environmental liabilities and financially material environmental costs.

- **Environmental accounting as an aspect of management accounting** serves business managers in making capital investment decision, costing determinations, process / product design decisions, performance evaluations, and a host of other forward looking business decisions. Thus, environmental accounting at this level refers to the use of data about environmental costs and performance in business decisions and operations.

In the same way, Environment Agency, UK has defined environmental accounting at these 3 levels. [8] It has summarised environmental accounting as, "The collection, analysis and assessment of environmental and financial performance data obtained from business management information systems, for example, environmental management and financial accounting systems...the taking of corrective management action to reduce environmental impacts and costs; plus, where appropriate, the external reporting of the environmental and financial benefits in verified corporate environmental reports or published annual reports and audited accounts." Exhibit 2.1 shows environmental accounting at these 3 levels.

Exhibit 2.1: Types of Environmental Accounting

Type of Environmental Accounting	Focus	Audience
National Income Accounting	Nation	External
Financial Accounting	Firm	External
Managerial or Management Accounting	Firm, Division, Faculty, Products, Lines, or System	Internal

Source: EPA, "An introduction to Environmental Accounting as a Business Management Tool: Key Concepts and Terms", US: EPA primer, Washington DC, (1995a), p.4.

The present study deals with environmental accounting and reporting at corporate level only. In the study, environmental accounting has been defined as, identification, measurement, recognition and disclosure of environmental costs, benefits, assets and liabilities.

2.2.2 Benefits of Environmental Accounting: As seen in the previous Section, environmental accounting relates to the identification, measurement, recognition and communication of environmental information to the interested parties. It is also popularly called as 'green accounting'. Conventional financial accounting does not deal with environmental issues separately. However, environment being an important factor in the establishment, survival and growth of an entity, there is ample justification for separate environmental accounting. In this sub-section, an attempt has been made to list some important benefits of environmental accounting.

The common accounting practice has been is to hide the environmental costs in two ways: firstly, by burying them in 'non-environmental' accounts and secondly by failing to link them to the activities that spawn them. As a result, managers are forced to take crucial business decisions like what products to manufacture, what techniques to use and what materials to purchase- without command of all relevant facts. But these days, because of changing regulations, increasing public concern and corporate commitments, it has become very important for a company to account for environmental costs. The major disadvantage resulting from following the conventional accounting model (with hidden environmental information) is that the organisation is deprived of many opportunities to increases profits, to use materials more efficiently and to protect the environment. The concept of environmental accounting would lead to better exploitation of these opportunities. The firms of any size and in any industrial sector can make use of environmental accounting information to obtain these gains.

Wilmhurst [9] identified seven benefits of environmental accounting:

- Provides to the firm, the better estimates of the true cost of producing a product. This improves pricing and hence, profitability.
- Allocates costs to the appropriate product, process, system or facility and thus, reveals costs to the responsible managers.
- Assists managers in targeting cost reduction, improving environmental quality and in reinforcing quality principles.
- Motivates staff to search for creative ways to reduce environmental costs.
- Encourages changes in processes to reduce waste, reduce resource use, recycle waste or identify market for waste.
- Increases employee awareness of occupational health and safety issues and
- Increases the likelihood of the company having a competitive advantage and greater customer acceptance of the firm's product or service.

Environmental accounting is useful not only for a company, but also for various stakeholders like customers, employees, government and society at large. From organisation's point of view, environmental accounting information leads to better decision-making, improved pricing and profitability. It helps in gaining competitive advantage by minimising adverse environmental impacts through improved designs, products or processes. It also assists in complying with legal requirements and discharging organisation's social responsibility.

From employees' point of view, environmental accounting not only increases their awareness of occupational health and safety issues, but also motivates them to search for creative ways to reduce environmental cost. Government can use environmental information published by various companies to ensure optimum allocation of scarce resources through changes in budgets and by other measures. International bodies (like International Monetary Fund, World Bank, and United Nations) also look for environmental information before giving assistance to a country. Because of the ethical investment movements in various countries, ethical investors (particularly institutional investors) require environmental information from companies. Last, but not the least, environmental accounting discloses unsound production processes, excessive use of natural and other scarce resources and thereby leads to conservation of energy, materials and other scarce resources and encourages their efficient optimum use. This is beneficial for society at large.

Thus, environmental accounting helps an organisation in reducing costs, improving profitability through better decision-making and enhancing image by meeting social responsibilities. It is also beneficial to various interest groups like shareholders, lenders, employees and government in making relevant investment and other decisions.

2.2.3 Environmental Accounting: Some Specific Issues: As discussed before, environmental information can be used not only by management for internal decision-making, but also by external stakeholders (including investors, creditors, government and society at large) to assess an enterprise's financial and environmental risks, and the impact of the company's environmental performance on its financial health. In this sub-section, an attempt has been made to discuss some of the specific issues relating to environmental accounting.

Before going into details of environmental accounting, a fundamental question that may be raised is: 'Is existing financial accounting framework adequate to take care of accounting and reporting on environmental issues?' In other words, 'Can traditional accounting framework deal with environmental costs, benefits, assets and liabilities?' As far as full cost accounting is concerned, it takes into account all the costs incurred by an enterprise including external environmental costs like consumption of non-renewable resources, damages to the environment, adverse impact on human health and environmental degradation. Here, the most important problem faced by accountants is, how to measure or value social costs and benefits (e.g. cost of destroying our resources or polluting the environment or social benefits of environmental protection measures). Existing financial accounting framework does not deal with these costs. It is difficult to measure these costs and benefits and hence, considerable research is required to be undertaken before developing suitable standards in this regard.

Regarding internal costs and benefits, there seems to be a general agreement among those developing standards that the basic concepts presented in various financial accounting frameworks are equally applicable to the environmental costs and liabilities, but additional disclosures may be required. [10] However, there are some specific accounting problems, which cannot be dealt with under the existing accounting framework (e.g. separation of environmental costs from other costs, capitalisation and expensing of environmental costs, measurement of environmental liabilities, disclosure of environmental benefits or treatment of pollution permits in the books of accounts). Though initiatives are being undertaken to facilitate collection of data pertaining to an enterprise's environmental costs and liabilities and to increase awareness of the financial implications of environmental issues, still no accounting standard or guideline specifically deals with these issues. Most environmental issues are not explicitly taken into account by the existing financial accounting standards and so the economic consequences of environmental impacts are sometimes underestimated. This reduces the incentive to improve eco-efficiency. A separate standard is, therefore, required to deal with specific environment related issues. In addition, some existing standards are required to be modified or revised so as to incorporate environmental considerations. The main purpose of specifically considering environmental issues in the accounting standards is to improve the usefulness of the information supplied to stakeholders. Some of the specific accounting issues related to environmental accounting are:

1. Identification of environmental costs.
2. Recognition of environmental costs: their expensing, capitalisation and amortization.
3. Recognition and measurement of environmental liabilities.
4. Environmental benefits.

1. Identification of Environmental Costs: Environmental costs are those costs that are incurred in complying with environmental laws or for protecting and conserving environment according to legal requirements or as per company's policy. "Environmental costs comprise the costs of steps taken or required to be taken, to manage the environmental impact of an enterprise's activity in an environmentally responsible manner, as well as other costs driven by the environmental objectives or requirements of the enterprise." [11]

Thus, environmental costs include voluntary as well as statutory costs incurred by an enterprise to reduce adverse impact of its activities on the environment. There are many ways to classify environmental costs. EPA has classified environmental costs into potentially hidden costs, contingent costs and image and relationship costs. Exhibit 2.2 shows some of the environmental costs incurred by a firm. It shows the following costs relating to the environment:

- **Conventional Costs:** These are costs of raw materials, utilities, capital goods and supplies. These costs are considered in cost accounting and capital budgeting, but are usually not considered as environmental costs. These costs and potential cost savings from them are generally overlooked in business decision-making.
- **Potentially Hidden costs:** These costs include upfront environmental costs which are incurred prior to the operation of a process, system or facility; regulatory and voluntary environmental costs incurred in operating that

process system or facility; and back-end environmental costs which will have to be incurred in the future. While management accounting systems may obscure upfront and current regulatory or voluntary costs, generally back-end costs are overlooked unless they are well documented.

- **Contingent Costs:** The contingent costs may or may not be incurred in future, e.g. costs of remedying and compensating for future accidental releases of contaminants into environment. These costs may not currently need to be recognised and therefore, may not receive adequate attention in internal forward-looking decisions.
- **Image and Relationship costs:** These costs are less tangible or intangible in nature, e.g., costs of annual environmental reports and community relations activities, costs incurred for awards programs. These costs themselves are not intangible, but the benefits that result from them often are.

Some of the above costs are clearly environmental costs, e.g., cost of purchasing pollution control equipment or fines and penalties paid for emissions above standards. But some costs are not clear environmental expenses, e.g., cost of purchasing energy efficient equipment. This is ordinary business expenditure, but additional outlay may be required to make it environment friendly. In the absence of a standardised definition of environmental expenses, it is left to the discretion of an enterprise to decide which expenditure would be included under environmental expenses. Enterprises experience great difficulty in separating environmental costs from other costs (particularly in case of capital expenditures) because in a number of cases, investments made to effect environmental improvements also result in expanded or improved production capacities. Bifurcation of expenses, in such cases, is an extremely difficult task. These costs can be bifurcated from other costs on different bases:

Exhibit 2.2: Examples of Environmental Costs Incurred by Firms

Regulatory	**Upfront**	**Voluntary**
Potentially Hidden Costs		
Notification	Site studies	Community relations
Reporting	Site preparation	outreach
Monitoring/testing	Permitting	Monitoring/testing
Studies/modelling	R & D	Training
Remediation	Engineering and	Audits
Recordkeeping	procurement	Qualifying suppliers
Plans	Installation	reports (e.g., annual)
Training		environmental reports)
Inspections	**Conventional Costs**	Insurance
Manifesting	Capital equipment	Planning
Labelling	Materials	Feasibility studies
Preparedness	Labour	Remediation
Protective equipment	Supplies	Recycling
Medical surveillance	Utilities	Environmental studies
Environmental	Structures	R & D
Insurance	Salvage value	Habitat and wetland
Financial assurance		protection
Pollution control	**Back-End**	Landscaping
Spill response	Closure/	Other environmental
Stormwater	decommissioning	projects
management	Disposal of inventory	Financial support to
Waste management	Post-closure care	environmental groups
Taxes/fees	Site survey	and/or researchers
Contingent Costs		
Future compliance costs	Remediation	Legal expenses
Penalties/fines	Property damage	Natural resource
Response to future	Personal injury	damages
releases	damage	Economic loss damages
Image and Relationship Costs		
Corporate image	Relationship with	Relationship with
Relationship with	Professional staff	lenders
Customers	Relationship with	Relationship with
Relationships with	workers	host communities
investors	Relationship with	Relationship with
Relationship with	suppliers	regulators.
insurers		

Source: EPA, "An Introduction to Environmental Accounting as a Business Management Tool: Key Concepts and Terms", US: EPA primer, (1995a), p. 9.

- Accounting standard or guideline
- Industry Guidelines
- Internal Mechanism

The most preferable basis is accounting standard or guidelines. If accounting standards are followed, it will reduce manipulations/distortion of accounts and will lead to standardised treatment in accounts. In the absence of any standard, industry guidelines can be used as a basis for bifurcation. [12] Companies can also use their internal mechanism for bifurcation. It gives discretion to the management to segregate expenses according to the need for information or requirements of the case.

"The success of environmental accounting does not depend on 'correctly' classifying all the costs incurred by an enterprise. It depends on making relevant information available to those who need or can use it." [13] Separating environmental costs from other costs or overheads and allocating them to appropriate product, process, system or facility helps in revealing these costs to managers, cost analysts, engineers or designers. It is necessary, not only for estimating correct production costs of different products or processes, but also helps managers target cost reduction activities that can also improve environmental quality.

Thus, the identification of environmental expenses provides a comprehensive picture of the efforts that have been undertaken by a business to protect environment. It also helps management in knowing true production costs, taking adequate steps to reduce environmental costs and thereby in increasing profitability.

2. Capitalisation and Expensing of Environmental Costs: The second challenge for accountants is, how to depict environmental outlays in financial statements when they are made, i.e., whether to capitalise them or charge them to revenue account and how to attribute them to different accounting periods. The broad basic criteria for capitalising such expenses ought to be the same as that followed in the financial accounting, with details being worked out for different types of pollution prevention measures. Some of the recommendations of the United Nations' Intergovernmental Working Group of Experts on International Standards of Accounting and Reporting (ISAR) are: [14]

A. Environmental Costs should be charged to income in the period in which they are identified, unless they meet the criteria for recognition as an asset.

B. If environmental costs meet the criteria for recognition as an asset, they should be capitalized. In other words, environmental costs should be capitalised, if they relate directly or indirectly, to future economic benefits that will flow to an enterprise through

- Increasing the capacity or improving the safety or efficiency of other assets owned by the enterprise;
- Reducing or preventing environmental contamination likely to occur as a result of future operations; or
- Conserving the environment.

Environmental costs that do not meet the criteria for recognition as an asset should be charged to Profit and Loss account. In other words, expenditures that result in the creation of a new asset or extend the asset's life, increase its existing capacity to reduce damaging influences on the environment or mitigate or prevent future contamination should normally be capitalised. Financial Accounting Standards Board's (FASB) Emerging Issues Task Force (EITF) also recommends capitalisation or expensing of certain costs incurred for environmental purposes on the same basis. [15]

C. When an environmental cost that is recognised as an asset is related to another asset, it should be included as an integral part of that asset, and should not be recognised separately.

D. When an environmental cost is capitalized and included as an integral part of another asset, the combined asset should be tested for impairment and where appropriate, written down to its recoverable amount. Similarly, capitalized environmental costs, recognised as a separate asset, should also be tested for impairment.

3. Identification and Measurement of Environmental Liabilities: These two issues are more problematic because environmental liabilities deserve special attention. Generally, future obligations are regarded as liabilities when there is a legal obligation to remedy an existing situation. On account of the inherent unforeseenability of future legislation, technological changes and the extent or nature of environmental clean up required, long term environmental liabilities are difficult to measure. Because of the uncertainties involved in environmental liabilities, in some countries, environmental liabilities are classified as provisions. But according to ISAR, the term environmental liabilities can be used to cover both kinds of costs- those which are certain, and those with some element uncertainty. ISAR, in its guidelines, provide: [16]

1. Environmental liabilities are obligations relating to environmental costs that are incurred by an enterprise and that meet the criteria for recognition as a liability.
2. A liability is a present obligation of the enterprise arising from past events, the settlement of which is expected to result in an outflow from the enterprise, of resources embodying economic benefits.
3. An environmental liability should normally be recognized when there is an obligation on the part of the enterprise to incur an environmental cost.
4. An obligation does not have to be legally enforceable for an environmental liability to be recognized. It could be a constructive obligation. [17]
5. Where there is difficulty in estimating an environmental liability, the best possible estimate in the range should be provided. In the rare situations, where an environmental liability cannot be estimated, disclosure should be made of the existence of such liability along with the reasons why an estimate cannot be made.

6. Expected recoveries from third parties should not be netted against environmental obligations. These items should be reported separately; the latter as a liability, the former as an asset.
7. With respect to the measurement of environmental liabilities that will be paid over a long period of time and that relate to future site restoration, closure and removal of long-lived assets, ISAR proposes a number of approaches. These include

- The *present value approach* which ISAR considers to be the preferred approach. Under this approach, the liability is measured at the present value of the estimated future expenditures that will be needed. The discount rate to be used should be risk free rate, such as the rate of government securities.
- The *current cost approach*, which ISAR considers to be an acceptable alternative. Under this approach, the liability should be measured at the estimated cost of performing the required activities in the current period.
- The *anticipated expenditure approach*, which ISAR does not support. Under this approach, anticipated expenditures over the life of the related operations are based on the undiscounted cash outflows that will eventually be required.

4. Recognition of Environmental Benefits: It is a common criticism of those advocating broader environmental disclosure that companies only present cost perspective and very rarely see (and report) the potential benefits resulting from incurring these environmental costs. Companies often argue that most of these benefits are not tangible and are expected only in future. But even realised financial benefits are not separately disclosed in the financial statements. Various surveys have also shown that while the accounting systems might often identify and indeed, emphasise the costs associated with environmental initiatives, there was no corresponding emphasis on the benefits (environmental, financial or otherwise). [18] This has the obvious effect of discouraging rather than encouraging environmental projects, performance and initiatives.

Eco-efficiency suggests that in addition to reporting environmental costs, a company should also report environmental benefits. In a given period, there are three types of benefits: income, current savings and cost avoidance. [19] *Income* refers to revenues that flow into organisation due to environmental actions like recycling paper, finding new applications for non- hazardous waste (e.g. using wood scraps to make wood chess pieces and boards) and increased sales due to an enhanced environmental image. *Cost avoidance* refers to ongoing savings that had been paid in prior years. *Current savings* refers to reduction in environmental costs achieved in the current year. By comparing benefits produced with environmental costs incurred in a given period, a type of environmental financial statement can be produced. Managers can use this statement to assess environmental performance of the organisation. This environmental financial statement could also form part of an environmental progress report that is provided to the shareholders on an annual basis. Thus, in order to assess cost-effectiveness of environmental protection measures undertaken, environmental benefits should be considered by a company. A formal cost-benefit analysis will help in that. It will also strengthen a company's commitment towards better environmental performance.

To summarise this Section, accounting of environmental information and its subsequent disclosure to various stakeholders is required to help them in making various financial and other relevant decisions. It can help an organisation in actually quantifying/measuring its environmental activities and performance. It will also help in proving that it is not just window-dressing, but the company is actually taking actions to protect and improve environment with tangible financial and other benefits. But in order to make environmental accounting successful, it is necessary that various accounting bodies should come up with some specific accounting standard or guidelines on environmental issues. A separate accounting standard and/or modifications in the existing standards, therefore, are required to deal with specific accounting issues relating to environment.

2.3 Environmental Reporting

Previous research has revealed that increasing public consciousness, high profile environmental disasters and stricter governmental regulations in some countries have resulted in a growing demand for corporate environmental reporting. As a large number of companies respond to these pressures by publicly disclosing their environmental performance in formal reports, demand increases on other companies to follow suit. Moreover, when a company adopts a proactive approach to environmental management, the desire to communicate the positive results to the concerned stakeholder groups also encourages preparation of environmental reports. As a result, since early 1990s, the number of companies reporting environmental information has increased substantially. But due to lack of accounting standards and guidelines on the issue, there are broad variations in the format and content of these reports. Hence, there is a need to provide some guidelines to promote consistency in content and quality of these reports.

This Section attempts to highlight some important issues relating to environmental reporting. The Section is further divided into 3 sub-sections. In sub-section 1, some important definitions of the term environmental reporting have been given. Sub-section 2 lists important benefits of disclosure of this information by a company. Finally, sub-section 3 presents the conceptual framework of environmental reporting.

2.3.1 Definition of Environmental Reporting: Environmental reporting is disclosure of information relating to environment in annual reports or by some other medium like in environmental policy statements or corporate environmental reports (CER). Some important definitions of the term environmental reporting are:

`"Environmental reporting covers the preparation and provision of information, by management, for the use of multiple stakeholder groups (internal or external), on the environmental status and performance of their company or organization. This information is most often provided in a separate environmental report, but it may (either as well or alternatively) be included in other forms of reporting (such as financial and social/ethical reporting)" [20]

`"Environmental reporting is the term now commonly used to describe the disclosure by an entity of environmentally related data, verified (audited) or not, regarding environmental risks, environmental impacts, policies, strategies, targets, costs, liabilities or environmental performance to those who have interest in such information; as an aid to enabling/enriching their relationship with the reporting entity via either

- the annual report and accounts package;
- a stand-alone corporate environmental performance report (CER);
- a site-centred environmental statement; or
- some other medium (e.g. staff newsletter, video, CD-ROM, internet site)." [21]

These days, environmental reporting is described either as a branch of the corporate governance tree, or as one aspect of the so-called 'triple bottom line'–whereby data on financial results, environmental performance and social impact are brought together in what might be termed as a sustainability report.

"Corporate environmental reporting should include environmental protection initiatives taken by the enterprise, the adverse impact of its production process and products on the environment both in quantitative and qualitative terms and its initiatives in process and product innovations in order to achieve sustainable growth." [22]

Thus, environmental reporting refers to the disclosure of environment related information by a company regarding environmental risks, impacts, policies, strategies, targets, costs and liabilities to the interested stakeholders through the annual report or some other medium like video, Internet or staff newsletter.

2.3.2 Costs and Benefits of Environmental Reporting: There are a large number of factors which drive a company to report on environmental issues like international standards, mandatory national requirements, competitive advantage, investment conditionality or other stakeholder pressures. Costs of environmental reporting are mostly direct like costs involved in publishing environmental reports or costs of environmental audits. But costs of non-reporting are mostly indirect like poor environmental profile vis-à-vis competitors. Exhibit 2.3 shows these costs of environmental reporting and non-reporting.

Exhibit 2.3: Costs of Reporting and Non-Reporting

Costs of reporting are mostly direct	Costs of not reporting are mostly indirect
• installing the appropriate environmental management systems • employing specialist staff/internal auditors • appointing external verifiers • publication/distribution cost/website design cost • there is also a potential "reporting rise" cost	• poor environmental profile vis-à-vis competitors • potential loss of markets/investors • loss/foregoing of other benefits of environmental reporting

Source: Ansari A.A., "Environmental Accounting and Reporting", The Indian Journal of Commerce, Vol. 54, No. 3, (July – September2001), p. 88.

The benefits of environmental reporting vary from company to company. Some of these benefits are:

1. **Improves Company's Reputation:** By demonstrating concern about environmental issues, a company can

distinguish itself from competitors and attract customers and suppliers. It enhances image of the company as a responsible employer and as a good corporate citizen. It gives competitive advantage to the company.

2. **Strengthens Company's Commitment:** Reporting environmental performance strengthens company's commitment to gain a higher level of environmental performance. Highlighting resource usage, waste and emission reduction or cost saving approach establishes environmental issues as a key policy/strategy element.
3. **Cost Savings:** An environmental report can not by itself reduce resource use or result in lower energy, water and material bill; but the preparation of such reports can focus the organisation's attention on such issues and can help it to identify possibilities for efficient saving. [23]
4. **Transparency and Communication:** Reporting can help convey the organisation's values and promotes two-way communication with its stakeholders. Environmental reports need to be comprehensive and actual work needs to be reported. It allows a company to demonstrate its improving environmental performance to pressure groups, the media and the general public.
5. **Staff Productivity and Quality:** People want to work for organisations that share their ethics and values. Organisations demonstrating and communicating environmental responsibilities are likely to recruit better people and are likely to have a more contended and productive work force.
6. **Risk Management:** Environmental risk is an integral part of risks facing every organisation. Reporting helps to identify such risks and where they could arise and thus, prevent damage to reputation from negative publicity on the issue. It helps in avoiding risk of action from Government or public opposition. Publishing the work done by a company also enables it to reassure investors/lenders as to environmental risks involved in

investment.

In nutshell, because of growing legal and social pressures, a large number of companies have started reporting on the environmental issues. There are many costs involved in environmental reporting. But at the same time, many benefits can be availed through it. It is difficult, but not impossible, for a company to undertake a formal cost- benefit analysis of the pros and cons of environmental reporting. It may show that there are a few 'real costs' and some very 'substantial benefits' accruing from the activity. [24]

2.3.3 Conceptual Framework for Environmental Reporting: "If corporate environmental reporting has to achieve the same degree of public acceptance as financial reporting, more attention needs to be paid to the various qualitative characteristics." [25]

The role of all the conceptual frameworks [26] is to standardise and underpin external corporate reporting and, thus, to give users greater confidence in the reporting process itself and the credibility of the information reported. [27] Preparers of environmental reports would like confirmation that their reports are effective and users of such reports (in particular an increasingly environmentally aware financial community) demand more consistency in the way(s), in which environmental issues and performance are measured and reported. A formal set of recognised reporting principles and standardised reporting frameworks would help overcome any perception that environmental reports lack credibility. A number of environmental reporting guidelines are available to (prospective and actual) preparers of environmental reports. The focus of these guidelines is mainly on 'what should be reported', rather than on the qualitative characteristics that would make these reports more useful (i.e. the how of reporting). In addition to these guidelines, there are emerging some national regulations concerning environmental reporting (e.g., in Denmark, Australia and the Netherlands). However, no accounting body (except FEE) has yet issued a conceptual framework for environmental reporting.

In this sub-section, the conceptual framework of environmental reporting has been presented. This framework mainly follows the pattern recommended by FEE.

1. Objectives of Environmental Reporting: The FEE Environmental Task Force defines the objectives of external environmental reports as being the provision of information about the environmental impact and operational performance of an entity that is useful to the relevant stakeholders in assessing their relationship with the reporting entity.

2. Users of Environmental Information: The International Accounting Standards Committee (IASC) conceptual framework [28] differentiates between seven user groups, but is based upon a primary assumption that the needs of financial investors are pre-eminent, and that if their needs are met, then the needs of most other users should be at least partially met. Research into the needs of users of environmental reports shows that the information needs of different environmental stakeholder groups can be substantially different. [29] Exhibit 2.4 shows information needs of various stakeholder groups.

ISAR has identified the following users of financial and environmental information: [30]

Financial Reporting:

- Present or Prospective Shareholders

Environmental Reporting:

- Shareholders/Potential owners
- Internal line managers, employees, competitors
- Lenders/Insurers
- Financial advisors/analysts
- Journalists/media
- Suppliers/customers
- Neighbour, local community
- Local and regulatory authorities (and national governments)
- Environmental organizations/NGOs.

Exhibit 2.4: Information Needs of Environmental Stakeholders

Investors: environmental provisions, liabilities, contingencies, expenditures, risk exposures and financial implications; environmental performance may impact on going concern status. Environmental and ethical probity and sustainability of the company.	**Employees:** Impact of business activities on non-renewable elements of the environment; information on the enterprise's environmental status; is the employer seen by local community groups/wider stakeholder group as a responsible company?	**Lenders:** direct financial implications or risks associated with environmental issues; quality of an entity's assets, its environmental management systems, its compliance record, its technological status and its market-place positioning.	**Suppliers and other trade Creditors:** Under-standing the environmental concerns and requirements of their major customers.

Contd...

Customers: Environmental liabilities and regulatory compliance issues; links with suppliers in terms of the environmental life-cycle; accreditation to a recognized Environmental Management System (e.g. EMAS or ISO 14001).	**Governments and their Agencies:** assurance of good environmental management assists in regulatory and policy decisions may affect the level and cost of monitoring and inspection. Environmental expenditure and regulatory (consent) compliance. Environmental reporting should reflect the response and pace of progress of entities operating at the micro level	**The Public:** as members of the local community, as customers, as employees, as shareholders. Are enterprises behaving as good neighbours in the local community? From a compliance perspective and also in respect of longer-term sustainability. Environmental performance may also affect longer term viability. Is the enterprise making a contribution to the local economy?

Source: Martin and others, 'An FEE Approach to Environmental Accounting', op. cit.

Thus, there are various users of environmental information having differing information needs. The company needs to recognise its relevant stakeholders and provide information on environmental issues accordingly.

3. Qualitative Characteristics: The building blocks of IASC conceptual framework includes 'underlying assumptions' and 'qualitative characteristics'. [31] Broadly speaking, the difference between an underlying assumption and a qualitative characteristic is that underlying assumptions have a more direct and significant effect on the absolute numbers reported. Qualitative characteristics are no less important for the overall credibility of reporting, but tend to address non-financial, evidential issues.

Reporting on the nature of environmental impacts means that in order to make the processes of measurement and reporting feasible and credible, certain underlying assumptions may need to be made about either the reporting enterprise or the data being reported on. [32] This is partly due to the fact that environmental impacts do not always naturally occur in the same time as the activity which causes them. Similarly, impacts do not always occur within the physical boundaries of the reporting enterprise. More importantly, there is a need to present data in as uniform manner as possible, if it is to be of any significant use to external shareholders.

Exhibit 2.5 reveals some important underlying assumptions and qualitative characteristics of environmental reporting.

The exhibit reveals that most of the underlying assumptions and qualitative characteristics of financial reporting are equally applicable to environmental reporting also (e.g. the entity assumption). A minimum requirement for environmental reporting is that the report itself (or any accompanying verification statement) clearly identifies the extent to which the entity is fully disclosing the significance and impact of its environmental activities. Similarly, liabilities for environmental costs (such as land remediation) will need to be recognised in the financial statements under the going concern concept, as long as there is either a legal or constructive obligation present. In the same way, for making environmental reports useful for external stakeholders, it must possess certain qualitative characteristics, e.g. it should be relevant, reliable and understandable. However, application of some characteristics is a bit complex in environmental reporting, e.g., the materiality concept. The application of materiality concept in case of environmental reporting is heavily dependent on the nature and circumstances of an item or event (as well as its scale). In particular, the carrying capacity of the receiving environment (such as availability of landfill capacity or background air pollution levels) is central to the issue of environmental materiality. Moreover, environmental reports may contain some information, which is neither objectively determined nor physically quantified (e.g., environmental impact of company's activities and long-term sustainability). Accordingly, some qualitative characteristics have to be modified.

Exhibit 2.5: Qualitative Characteristics of Environmental Reporting

Underlying assumptions of environmental reporting	Qualitative characteristics important to public environmental reporting
• The entity assumption • The accruals basis of accounting • The 'going concern' assumption • The 'precautionary' principle • The concept of materiality of significance	• Relevance • Reliability • Understandability • Neutrality • Completeness • Prudence • Comparability • Timelines • Verifiability

Source: FEE, 'Discussion Paper Towards a Generally Accepted Framework for Environmental Reporting', (1999), op. cit.

Thus, most of the assumptions and concepts of financial reporting are applicable in case of environmental reporting also. But some modifications according to quality, need or use of the information have to be made. The standard setters should make adequate provisions accordingly while setting conceptual framework for environmental reporting.

4. Tailoring Environmental Reports to Users' Needs: Companies in most of the countries use different approaches to present their environmental disclosures, as the environmental reporting is not regulated in these countries. Disclosure normally falls in 3 categories:

- Descriptive and Qualitative reporting
- Quantitative environmental accounts
- Financial environmental reporting

There is no single correct way of publishing or communicating environmental performance data. A reporting system for environmental impacts can use either one or some combination of technologies–narrative, numerical and graphic (charts, graphs, photographs, computer displays) in order to disclose relevant information to users. [33]

The environmental information can be reported in the annual report or separate environmental statement or by some other medium like websites and newsletters. A company should identify its relevant stakeholders and publish the information accordingly. For example, employee related information can be reported through employee newsletter; customer directed reports for customers; community directed site reports or financial information in annual reports.

To conclude, environmental reporting is now a major part of corporate reporting. In the 1990s, there was significant growth in environmental reporting by large companies. But there is an urgent need to improve the conceptual underpinning and thus, the quality of external environmental reporting itself. Conceptual underpinning requires the development of a conceptual framework of environmental reporting containing objectives, general assumptions, qualitative characteristics and guidelines for companies. A more disciplined approach, therefore, is required for reporting environmental information.

2.4 Environmental Audit

Reporting of environmental information to external stakeholders gives rise to the need for verifying these statements i.e. environmental audit. Environmental auditing originated in the United States in the 1970s, as a way of checking whether a company was complying with a multitude of new environmental laws and regulations. Now days, it is used as an extremely valuable tool for assessing a company's environmental management systems, policies and procedures. In this Section, an attempt has been made to discuss some issues relating to environmental audit. This Section is divided into 3 sub-sections. Sub-section 1 presents some important definitions of environmental auditing. Sub-section 2 gives objectives of environmental auditing. Finally, sub-section 3 takes up some general issues relating to audit of environmental statements.

2.4.1 Definition of Environmental Audit: Environmental audit refers to verification and assessment of environmental measures and performance of an organisation. According to Arthur D. Little, "Environmental audit is a systematic and objective method of reviewing environmental management system and controls and verifying that environmental standards, regulations, company and industry standards are being met." [34]

This is a narrow definition of environmental audit, concentrating mainly on compliance part. Some more comprehensive definitions of environmental audit are:

"A management tool comprising a systematic, documented, periodic and objective evaluation of the performance of the organisation, management and equipment designed to protect the environment, with the aim of facilitating management control of environmental practices and assessing compliance with companies' policies, including observance of the existing regulatory requirements." [35]

"Environmental auditing is a systematic examination of the interactions between any business operations and its surroundings. This includes all emissions to air, land and water; legal constraints; the effects on the neighbouring community, landscape and ecology; the public's perception of the operating company in the local area. Environmental audit does not just check compliance with legislation, nor is it a 'green washing' public relations exercise... Rather it is a total strategic approach to the organisation's activities." [36]

These definitions show that environmental audit is a broad term including audit of environmental system, policies, programmes and performance of an organisation with a view to decrease the environmental liabilities and risks and to ensure compliance with environmental regulations. However, the present study concentrates on environmental audit only as audit of environmental information given by a company in the annual report, environmental account or statements prepared by it.

It has been observed in many studies on environmental reporting that the information communicated by a concern should be useful for external stakeholders in decision-making (see, for example, Jones, 2000; Deegan and Rankin, 1999; Moneva and Llena, 2000). For this, it is necessary that only relevant and reliable information is provided to them. This gives rise to the need for audit of environmental reports prepared by a concern. Environmental information audit (EIA) is verification of environmental information given by a company in the annual report, environmental accounts or statements prepared by it. This type of audit is required for establishing the credibility and integrity of environmental accounts whether physical or monetary, and the validity of statements, reports, and other documentation. Thus, in this sense, EIA may be defined as a systematic process of objectively obtaining and evaluating the evidence relating to the performance of an organisation as reflected in the environmental statements and examining and reporting the relevant quantitative and qualitative information to the interested stakeholders.

2.4.2 Objectives of Environmental Audit: Environmental audit, in general, seeks to obtain following goals: [37]

- to examine the compliance of environmental norms issued by the regulatory bodies.
- to assess and verify the environmental compliance with the corporate requirements.
- to evaluate and report key environmental performance like pollution control measures, waste management etc.
- to ensure quality improvement in every sphere of corporate life.
- to minimise wastage, losses and thereby resulting in cost reduction.
- to generate awareness and consciousness about environmental issues amongst all levels of employees in the organisation.
- to examine the efficient utilisation of materials and reduction in consumption on energy and more importantly, to develop a rational basis for utilizing environmental resources.
- to protect the corporation from potential liabilities.

To summarise, there are 3 main objectives of environmental audit: Compliance with regulatory codes; assistance in acquisition and disposal valuations; and corporate development towards green mission.

2.4.3 Specific Issues Relating to Environmental Audit: Whether a company's often repeated assertions of commitment to environmental protection have actually been translated into action, can only be ascertained when some independent verifier authenticates its published environmental reporting data. This kind of audit is very important to assure users of these statements about accuracy and validity of assertions made in these statements. The auditor has to check the statements relating to environmental performance of an organisation. He is also required to assess contingent liabilities including environmental risks. Exhibit 2.6 presents some areas where the statutory audit needs to reflect environmental considerations.

Exhibit 2.6: Some Areas in Which the Statutory Financial Statements and Audit Will Need to Reflect Environmental Considerations

- **Contingent liabilities:** especially on contaminated land, oil spills and unauthorized emissions.
- **Provisions:** especially for remediation, abandonment and decommissioning costs (e.g. the oil and gas company practice), but also waste disposal and recycling commitments as well as potential catch-up insurance and legal costs.
- **Reserves:** especially for catastrophes
- **Valuation of fixed assets**: especially land and buildings.
- **Depreciation policy:** to recognize, for example, shorter life of productive assets under rising BATNEEC considerations.
- **Additional capital costs associated with productive assets:** especially the need to incur additional costs to bring existing plant within current standards.
- **Obsolete inventory and inventory costs:** including stock made obsolete through environmental concerns, storage and disposal costs of environmentally malign materials and recycling commitments.

Source: Gray R. and Bebbington J., *Accounting for the Environment,* Sage Publications, London, 2nd ed., (2001), op. cit., p. 234.

1. External vs. Internal Auditors: Environmental audit can be conducted by internal auditors or by external independent auditors. Internal auditors as a part of management are interested in reviewing compliance with environmental regulations and statutes; determining the propriety of the accounting for environmental issues and ensuring that proper disclosure is being made. They evaluate the internal controls to see that these are in place to keep environmental problems at a minimum, to make efficient use of resources and to keep wastes and pollution under control. Thus, internal audit functions can cover many areas like compliance audits, energy audit, EMS audit; supplier audits, insurance audits, pollution prevention audits, and product and process audit.

External auditors are mainly interested in ensuring that financial statements depict a true and fair view of a company's trading results and financial position. For this, they must also review some of the elements of the organisation's environmental operations. They must be familiar with the environmental aspects of reviewing assets and liabilities to determine

- that their valuation is proper.
- that the contamination has not reduced the carrying value of the assets and
- that the expending and capitalisation of environmental costs have been recorded properly.

They will also see that financial statements reflect true assets and liabilities of the organisation. They either by themselves or with environmental experts, determine that the organisation is complying with governmental regulations in handling of emission of pollutants, the disposal of contamination and waste, and the detoxification of previously contaminated assets. In addition, the external auditors must determine whether the client is exposed to the risk of incurring liabilities or has actually incurred liabilities resulting from the contamination of recently acquired assets. The external auditor must also ensure that proper disclosure, in compliance with legal requirements has been made.

Since environmental audit is a diverse activity, in large organisations, it is necessary that an audit team should conduct it, instead of an individual. The audit team must be qualified for the operation. In addition to persons with audit ability, the team should have available to them qualified environmental engineers on a full time basis or as advisors when needed. The audit team should have legal expertise available as and when required.

2. Verification Standards: For verification, it is necessary to have standards against which information has to be verified. Appropriate verification standards for verifying environmental information are slowly beginning to emerge. The International Auditing Practices Committee (IAPC) and Environmental Task Force of FEE are both in the process of developing standards for the audit of environmental reports. [38]

ISO 14000 is a methodology for environmental management. It provides necessary requirements and recommendations for any organisation to develop and implement a cost effective system of management by introducing environmental auditing. ISO-14000 series covers environmental audit, audit procedures, audit criteria, audit management and environmental reviews, site assessment, labelling performance evaluation and life cycle analysis. Hence, ISO-14000 also provides basis for environmental audit. [39]

To summarise, environmental audit is an independent evaluation of policy and practices, systems, procedures and performance of a company relating to environment. There are many objectives of environmental audit like to evaluate efficiency and efficacy of resource utilisation (e.g. energy, material and other scarce resources), to identify areas of environmental risk, environmental liabilities and problems in complying with regulatory requirements. One of the very important areas of environmental audit is verification of environmental information produced by a company. The present study mainly concentrates on this part of environmental audit and thus, ignores other technical areas relating to environmental audit.

Summary

In this Chapter, various concepts and theoretical

considerations relating to EAR have been examined. It has been observed that the term environmental accounting can be explained at different levels. National level environmental accounting is a method of keeping accounts of natural resources and other environmental elements. It involves valuation of natural resources, measuring income therefrom, keeping records of costs relating to them, estimating them and providing depreciations on them. On the other hand, at corporate level, environmental accounting refers to the measurement and communication of information on the environmental responsibility performance of an entity to the interested parties. It is a part of corporate social responsibility and attempts to evaluate impact of organisational activities on environmental resources. While environmental financial accounting deals with collection, measurement and reporting of environmental information for external reporting purposes, environmental management accounting is concerned with presenting environmental data to management for internal decision-making.

Accounting of environmental information and its subsequent disclosure to various stakeholders is required to help them in making various financial and other relevant decisions. It can help an organisation in actually quantifying/measuring its environmental activities and performance. It will also help in proving that it is not just window-dressing, but the company is actually taking actions to protect and improve environment with tangible financial and other benefits. As environmental issues and their related costs grow in rupee size and in public awareness, accounting profession must be prepared to incorporate their impact into company's financial disclosures and annual reports. There is an urgent need for further work on how to integrate non-financial environmental information into financial accounting systems. There are a number of alternative valuation methods and useful tools that are now available in many countries for business to use in their environmental accounting. By considering social and environmental externalities an organization can start to assess its long-term sustainability. Also by helping to reduce the use of and the expenditure on natural resources and the cost and impacts of disposing of waste- the application of environmental accounting might help increase the profitability and productivity of both public and private sector organizations.

In nutshell, environmental issues have increased in importance in past decades. This trend is expected to continue in future and thus, accountants and managers need to be aware of environmental issues and to consider their influence upon both internal and external reporting. But in order to make environmental accounting successful, it is necessary that various accounting bodies should come up with some specific accounting standard or guidelines on environmental issues. A separate accounting standard and/or modifications in the existing standards, therefore, are required to deal with specific accounting issues relating to environment. Moreover, in order to provide relevant, reliable and comparable information to the interested stakeholders it is necessary that only duly audited information is provided to them. This gives rise to the need for independent environmental information audit.

Notes

1. The Oxford Concise Dictionary, Oxford University Press, Delhi, 9th ed., (1995).
2. The Macmillan Compact Encyclopaedia, Pan Macmillan Ltd., 1994 edition.
3. Environment (Protection) Act, 1986 of India.
4. Bhadra B., 'ISO 14001 and Environmental Cost Accounting- An Introspection', in Pramanik A.K., *Environmental Accounting and Reporting*, Deep and Deep Publications, New Delhi, (2003), p.467.
5. Oza H.S., 'Corporate Environmental Responsibility – Legal Aspects, Accounting and Auditing', *Ibid,* p. 90.
6. Gray R., Owen D., and Maunders K., *Corporate Social Reporting- Accounting and Accountability,* Prentice Hall International, London, (1987), p.60.

7. Environment Protection Agency (EPA), '*An Introduction to Environmental Accounting as a Business Management Tool: Key Concepts and Terms*', US EPA primer, Washington DC: EPA, (1995a), p.28.
8. Environment agency, '*Environmental Accounting – Useful Links*', Source: http:/environment. agency.gov.UK/business, last accessed on 13-08-02.
9. Wilmhurst T., '*A Review of Environmental Accounting Issues for Australian Accountants*', cited in a conference paper, prepared for the Australian Chamber of Manufacturers, Environmental Accounting and Reporting Conference, ASCPA, (19th June, 1997), See http://www.ea.gov.au., last accessed on 29-8-04.
10. UN-ICAI Seminar Report, '*Environmental Accounting and Reporting*', conducted at New Delhi, (Aug. 2-4, 1999), p. 5.2.
11. *Ibid.*, p. 5.7.
12. Many countries like Germany and Japan have industrial guidelines for this purpose. See Rathore S., *International Accounting,* Chapter 17- Accounting for Environmental Protection Measures, Prentice Hall of India Pvt. Ltd., New Delhi, (1996), p. 421.
13. EPA, *op. cit.*, p. 12.
14. UN-ICAI Seminar Report, *op. cit.*, pp. 5.7-5.8.
15. Pramanik A.K., *op. cit.*, p. 112.
16. Abu-Ghazaleh T., "Financial Services and the Environment: the Accounting Profession", *UNEP Industry and Environment,* (Jan.-March, 1999), p. 33.
17. A constructive obligation is one that can be created, inferred or constructed from the facts in a particular situation, rather than being legally based or .that arises from ethical or moral considerations and that an enterprise has little or no discretion to avoid.
18. Gray R. and Bebbington J., *Accounting for the Environment*, Sage Publications, London, 2nd ed., (2001), p.162.
19. Hanson D.R. and Mowen M.M., *Cost Management Accounting and Control,* Chapter 17 – Environmental Cost Management, South Western College Publishers, Cincinnati, USA, 3rd edition, (2000), pp.665-698.
20. Federation Des Experts Comptables Europeens (FEE), '*Discussion Paper Towards a Generally Accepted Framework*

for Environmental Reporting', (1999), Source: http://www.fee:brussess. fee.be/publications/main.htm.

21. UN-ICAI Seminar Report, *op. cit..*, p. 62.
22. Guidelines issued on Global Reporting Initiative (GRI), Source: www.globalreporting.org.
23. Evidence from the UK Govt.'s Envirowise programme suggest that a company that has made no previous efforts to improve its environmental management can typically reduce both water and energy costs by 15 to 20 percent with no cost or low cost measures, Source: www.Envirowise.gov.UK.
24. Gray R. and Bebbingten T., *op. cit.*, p. 255.
25. FEE (1999), *op. cit.*
26. The Financial Accounting Standards Board (FASB) USA, has defined a conceptual framework as a constitution, a coherent system of interrelated objectives and fundamentals that can lead to consistent standards and that prescribes the nature, function and limits of financial accounting and financial statements. See, FASB, *Conceptual Framework for Financial Accounting and Reporting*, (1976).
27. The objective of the conceptual framework is to help produce a body of standards that is more internally consistent than the ad hoc approach, thereby, enhancing the credibility of accounting information. See, Lal J., *Accounting Theory*, Himalaya Publishing House, Mumbai, 2nd ed., (2001), pp. 295-297.
28. IASC, "A Framework for the Preparation and Presentation of Financial Statements", *The Chartered Accountant*, (Aug. 1989), pp. 124-133. Also see ICAI's Framework in The Chartered Accountant, (July 2000), pp. 45-46.
29. Martin and others, "*An FEE Approach to Environmental Accounting*", Source: http://www.academy.com, last assessed on 10-7-2004.
30. UN-ICAI Seminar Report, *op. cit.*, p.6.12.
31. IASC, *op. cit.*
32. FEE (1999), *op. cit.*
33. Robert W. Lamson, 'Corporate Accounting for Environmental Effects', in Dierker M. and Bauer R.A. *Corporate Social Accounting*, Praeger Publishers, N. York, U.S.A., (1976), pp. 231-235.
34. Singh, Gurdeep, 'Overview of Environmental Auditing', in Kumar A. and Goel P.K., *Industry, Environment and Pollution*,

ABD Publishers, Jaipur, (2000), pp. 390-395.
35. Jariwala C.H., 'Environmental Auditing' in Pramanik A.K. *op. cit.*, p. 146.
36. Paramasivam Thirumoorthy, "Taking Professional Care of the Nature: The Environmental Audit", in *The Chartered Accountant*, (Aug. 2002), p. 148.
37. Charumathi B., "Added Accountability of the Corporate Sector for Environmental Concerns", *Journal of Accounting and Finance*, Vol. 12, No. 2, (Sept. 1998), p.8.
38. UN – ICAI Seminar Report, *op. cit.*, p.5.24.
39. Pramanik A.K., *op. cit.*, p. 61.

3

Environmental Accounting and Reporting: Regulatory Framework

The previous Chapter has shown that EAR is now an important part of corporate accounting and reporting practices. It has been increasingly realized that adopting an environmental accounting system helps the managers in making various business decisions (like capital budgeting, product and process design). When this information is disclosed to external stakeholders, they can use it in making relevant investment or other decisions. In this Chapter, an attempt has been made to summarise recent laws and guidelines relating to EAR in some selected countries. Section 3.1 gives regulatory framework on EAR in some selected countries. Section 3.2 examines the need for comprehensive EAR laws in India and Section 3.3 finally, summarises the Chapter.

3.1 Recent Ear Laws in Some Countries

In Chapter II, some specific problems relating to EAR were discussed. Guidelines regarding these issues have been issued by many organisations such as International Chamber of Commerce, The Japanese Industry Association, Financial Accounting Standard Board's (FASB) Emerging Issues Task Force, Canadian Institute of Chartered Accountants, United Nations Inter-Governmental Working Group of Experts on International Standards of Accounting and Reporting (ISAR). [1] But, unfortunately till date, no accounting standard has been issued for treatment of these specific problems relating to environmental accounting. As regards environmental reporting also, different organisations like FASB, FEE, GRI, ISAR etc. have issued guidelines but these guidelines are only advisory in nature and not mandatory. This Section presents some of the recent regulations and guidelines on EAR issued by various organisations from different countries.

The development of environment, health and safety policy statements has a long history in USA. [2] In this country, the accounting standards do not comprehensively address environmental accounting issues. However, there are a number of rules on reporting and recording of costs and liabilities that pertain to environmental matters.

The Financial Accounting Standard Board (FASB) came up with FASB Statement No. 5 which indirectly deals with some environmental issues. [3] This standard relates to "Accounting for contingencies" and provides guidelines for recognising losses when an existing condition, situation or set of circumstances involve uncertainty. An interpretation of the standard, "Reasonable estimation of the amount of a loss", also provides guidance on recognition and disclosure of contingencies including environment contingencies. It requires entities to recognize losses when it is "probable" that asset is impaired, the loss is "reasonably estimable", and the liability is material. FAS 143 was developed in 2001 to provide uniformity in how companies evaluate asset retirement obligations (ARO). FAS 143 applies to "legal obligations", associated with the retirement of tangible long-lived assets such as property and equipment. FAS 143 rejected the "probable and reasonably estimable" approach of FAS 5 in favour of a fair value measurement (FVM) when calculating AROs.

Since late 1990s, the Emerging Issues Task Force (EITF) of FASB has started examining some more accounting issues relating to environmental problems. An EITF report, "Accounting for the costs of Asbestos Removal", considers whether this cost should be capitalised or treated as an expense. Another EITF report, "Accounting for Environmental Liabilities" addresses measurement of loss when recoveries are possible and whether the liability should be measured or discounted or undiscounted basis. EITF report, "Capitalisation of costs to treat environmental contamination" also addresses the issue of capitalisation.

The current environmental disclosure requirements of Securities and Exchange Commission (SEC), USA, are shown in Exhibit 3.1. US public companies are required to comply with these requirements. SEC requires management to disclose in the "description of business" section of the annual report, the impact of material efforts that compliance with governmental provisions regulating the discharge of materials into the environment or otherwise relating to the protection of the environment, may have upon capital expenditures, earnings or the competitive position of the enterprise. [4] The firm is also required to disclose any material estimated capital expenditures for environmental control facilities for the remainder of the current fiscal year and its succeeding years and for such periods as the enterprise might deem material.

Exhibit 3.1: Environmental Disclosure Requirements of SEC, USA

- Effects of compliance with environmental laws which may have on capital expenditure, earnings and competitive position.
- Disclosure of pending or contemplated administrative or judicial proceedings, including those arising under environmental laws.

Source: Johnson L.T., "Research on Environmental Reporting", Accounting Horizons, 7(3), (Sept. 1993), pp. 118-123.

There are number of other countries that require companies to report on environmental and social issues. [5] A summary of environmental reporting requirements in some of these countries is given in Exhibit 3.2.

As depicted in Exhibit 3.2, in Australia, Corporations Law (section 299 [1f]) was introduced in 1999 requiring companies to provide details of the entity's performance in relation to environmental regulations in the Director's report. On 1 July, 2004, the Corporate Law Economic Reform Program (Audit

Reform & Corporate Disclosure) Bill 2003 (CLERP 9), extended this to the operations and financial position of the entity and its business strategies and prospects (Section 99A [1]).

Exhibit 3.2: Mandatory Environmental Reporting in Some Countries

European Union
• Based on the **Integrated Pollution Prevention and Control Directive (IPPC)**, member states will be required to register emission data from large companies (so called IPPC installations) and report this data to the commission.
Australia
• **Corporations Law Section 299(1) (f)** introduced in 1999 and requires companies that prepare a director's report to provide details of the entity's performance in relation to environmental regulations. • **Financial Services Reform Act** was introduced in March 2002 and requires fund managers and financial product providers to state "the extent to which labour standards or environmental, social or ethical considerations are taken into account in the selection, retention or realization of the investment." • **National pollutant Inventory** requires industrial companies to report emissions and inventories for specific substances and fuel to regulatory authorities for inclusion in a public database.
Belgium
• Article 4.1.8 of **VLAREM II** (1995) stipulates that certain companies have to issue an annual environmental report. Companies subject to this obligation are listed in Appendix 1 of VLAREM I and mainly concern Class I and Class II hazardous activities.
Canada
• The **Securities Commission** requires public companies to report the current and future financial or operational effects of environment protection requirements in an Annual Information Form.

Contd...

Denmark
• The **Law on Annual Accounts**, introduced in 2001, requires reporting on intellectual capital resources and environmental aspects in the management report, if it is material to providing a true and fair view of the company's financial position. • The **Law on Green Accounts** and statutory order from the Ministry of Environment and Energy has been applicable since 1996 and obligates certain listed activities/ companies to draw up green accounts.
France
• **"Law No. 2001-420 related to new economic regulations (Art. 116)"** *(May 2001)*, environmental and social reporting is mandatory for publicly-quoted companies, starting with data for the 2002 financial year.
Norway
• The **Accounting Act** (1999) requires that all companies include more detailed environmental information in Director's reports in annual financial reports from 1999 onwards
Sweden
• **Amendment to the Annual Accounts Act** applicable since 1999. Companies that are required to have environmental permits or must notify the environmental authorities, have an obligation to include a brief disclosure of environmental information in the board of director's report
The Netherlands
• The **Environmental Protection Act** (1997) includes a section on environmental reporting. This section indicates which companies are required to report on environmental matters. To date, over 250 companies have each published two reports a year: one public report and one governmental report.
USA
• In U.S.A., companies with over ten full time employees are required to submit data on emission of specified toxic chemicals to the EPA under the Toxic Release Inventory (TRI). In addition, the Securities and Exchange Commission requires disclosure on legislative compliance, judicial proceedings and liabilities relating to the environment in Form K-10.

Source: KPMG, *'KPMG International Survey of Corporate Sustainability Reporting'*, (2002).

Financial Services Reform Act 2001 commenced in March 2002 requires fund managers and financial product providers to state "the extent to which labour standards or environmental, social or ethical considerations are taken into account in the selection, retention or realization of the investment." In addition, National Pollutant Inventory requires industrial companies to report emissions and inventories for specific substances and fuel to regulatory authorities for inclusion in a public database. Australian Securities and Investments Commission's (ASIC) Disclosure Guidelines ask product issuers for disclosure about labour standards or environmental, social and ethical considerations in Product Disclosure Statements (PDS). Triple Bottom Line Reporting in Australia is a guide to reporting against environmental indicators consistent with the Guidelines of the Global Reporting Initiative (GRI).

In Canada, the Canadian Institute of Chartered Accountants (CICA) carried out a research study on accounting and reporting on environment in 1993, entitled "Environmental Costs and Liabilities: Accounting and Financial Reporting issues." [6] The objective of the study was to identify what constitutes environmental issues and environmental protection measures and the extent to which current accounting standards are an impediment to entities undertaking expenditure relating to the improvement and protection of the environment. The report went on to suggest ways of recognizing and recording environment related expenditure. In 1994 the CICA [7] published the study to guide the organisations on the various matters that should be taken into consideration while reporting on environmental performance. More recently, CICA approved an accounting standard on remediation costs, which has been passed by the Canadian Standards Committee. [8] The Securities Commission in Canada requires public companies to report the current and future financial or operational effects of environment protection requirements in an Annual Information

Form. In addition, the Bank Act requires banks and other financial institutions with equity of USD 1 billion or more to publish an annual statement describing their contributions to the Canadian economy and society.

In UK, the Institute of Chartered Accountants of England and Wales (ICAEW) in 1992 published a report on Environmental Disclosure. [9] The report was titled, "Business Accountancy and Environment: A Policy and Research Agenda." This report enlists the environmental disclosure requirements that companies should make as part of their annual reporting cycle. The Department of Institutional Investors (DII) and Department of Environment (DoE), to promote a dialogue on environmental issues and to help mobilise the business community in demonstrating good environmental practices and management, set up the Advisory Committee on Business and Environment (ACBE) in 1991. ACBE has also recommended disclosure of some environmental information by companies in UK.

The UK Financial Reporting Standard FRS-12 (Provisions, contingent liabilities and contingent assets) issued by Accounting Standards Board (ASB) in 1998, specifically looks at the accounting procedures required for dealing with issues, such as contaminated land. In addition, FRS-10 (Goodwill and intangible assets); FRS 11 (Impairment of fixed assets and goodwill); FRS 15 (Tangible fixed assets) and FRS 18 (Accounting policies) may also be helpful for organisations in implementing environmental accounting policies and procedures. [10] The Operating and Financial Review (OFR) will be a legal requirement for all UK listed companies to provide a narrative within their Annual Report on the company's strategies, performance, future plans and key risks which may include ethical, social, environmental, brand and reputational risks. The Combined Code as part of the Financial Services Authority's listing requirements requires organizations to report on governance and internal controls, which cover, among other things, material non-financial issues. [11]

In France, a concept of "Ecological Balance Sheet" has been developed, which concerns the relationship between an enterprise and the environment. [12] The Ecological Balance Sheet contains information about the acquisition and utilization of equipment used to reduce pollution, recycled by products and reduced energy and raw materials consumption. Under *"Law No. 2001-420" (May 2001)*, environmental and social reporting is mandatory for publicly quoted companies, starting with data for the 2002 financial year.

Bulgaria introduced environmental accounting in 1992 through the Accountancy Act. [13] This Act requires completion of two types of statistical forms to produce a single source of accounting and statistical information. All companies complete a form entitled "Environmental Protection Costs", which contains details of expenditure for environmental purposes on non-current assets such as, construction, rehabilitation and preservation cost of sites, the upgrading of equipment and intangible assets such as software and patents. The cost of maintaining these "environmentally motivated assets" is also reported.

In Spain, accounting standards are issued by ICAC (Instituto de Contabilidady Auditoria de Centas), which is a governmental agency. It issued an accounting standard for electricity companies in 1998, "Note 19th" of the standard includes information on the environment particularly in respect of environmental investments, environmental expenses, provisions for environmental issues and contingent liabilities. [14] The National Accounting Plan for the Electricity Sector specifies environmental issues in more detail.

In Japan, the Law of promotion of environmentally conscious business activities requires "specified entities", to publish an environmental report every year. The Pollutant Release and Transfer Register (PRTR) Law concerns reporting of releases to the environment of specific chemical substances. The Finnish Accounting Act requires companies to include material non-financial issues in their directors' report of the annual/financial report.

In Korea, the Korean Securities Exchange Commission (KSEC) enacted a provision in 1996, in the Corporate Accounting Standards (CAS) which requires the inclusion of environmental information in the form of accompanying footnotes to the corporate financial report. [15] Article 90 of the amended CAS recommends the voluntary disclosure of the following information in the footnotes attached to the financial report, "The Company's environmental standards and policy; safety and accident related matters, environment related investments; consumption of resources and energy; and matters related to occurrence and treatment of by-products and scraps."

In many other countries like South Africa, Italy, Poland, Malaysia, Bangladesh and Sri Lanka, there is no extensive statutory requirement for companies to disclose environmental information in annual reports. Though some of these countries have extensive environmental legislations and companies are required to furnish some environmental information to authorities on regular basis, but no accounting and reporting requirements to general public are there.

Since the early 1970's, business organizations in India have been subject to various environmental laws and regulations, the Environment (Protection)Act, 1986 being the most significant step in this direction. In order to control emissions from production processes and to protect environment, stringent requirements as to production processes and equipment use have been imposed in major polluting industries. A gazette notification on environmental audit was issued by Ministry of Environment and Forest (MoEF) on 3.3.1992, requiring submission of an Environmental Statement to the Pollution Control Board (PCB). [16] It is applicable to any person carrying on an industry, operation or process requiring consent from regulatory authorities to operate under any law relating to environment. In this environmental statement, the concerned industry is required to provide information on water and raw material consumption; pollution generated; nature of hazardous wastes and solid wastes and the disposal practices; impact of pollution control measures and conservation of natural resources. Companies are also required to publish information on Energy Conservation (steps taken for this and benefits derived therefrom) in the annexure to Director's Report under Sec. 217(1) (e) of Companies Amendment Act, 1988. Moreover, Companies Bill 1993 and 1997 proposed the amendment of Section 173 to require the board of directors to report the measures taken for protection of environment in such manner as may be prescribed by the authorities. [17]

Apart from these national regulations and guidelines, different types of supra-national bodies have also shown interest in environmental reporting and some of these have issued guidelines in this connection. The most significant and long standing initiative in the development of environmental reporting within conventional financial accounting framework has come from the *United Nations Inter Governmental Group of Experts on International Standards on Accounting and reporting (ISAR)*. Starting from 1989, ISAR has issued a number of recommendations on EAR. [18]

Significant developments of *European Union (EU)* regulation that had implication for environmental reporting began in the early 1990s. The commission in its Fifth Action Programme on the Environment, entitled "Towards sustainability" in 1992, indicated a number of environmental disclosures (like disclosure of environmental policies; activities and efforts; environmental expenses incurred during the year and environmental risks). The EU Eco-Management and Audit Scheme (EMAS, 1995) is a management tool for companies and other organizations to evaluate report and improve their environmental performance. In 2001, the European Commission adopted a recommendation on recognition, measurement and disclosure of environmental issues in the annual accounts and annual reports of EU companies. [19] The recommendation clarified existing EU accounting rules and provides guidance to improve the quality, transparency and comparability of environmental data available in company's annual accounts and reports. Exhibit 3.3 summarises the purpose of this recommendation.

Exhibit 3.3: EU Commission: Recommendation on Environmental Issues in Company's Annual Accounts and Reports

The recommendation

- Provides guidance on how to apply the provisions of the existing Accounting Directives as regards environmental information,
- Suggests closer coordination of separate environmental reports, statements in annual accounts and annual reports, so as to reduce inconsistencies,
- Indicates that relevant transparent disclosures should be incorporated into company's annual accounts and annual reports in a way that complements the more detailed separate environmental reports.

Source: http://europa.eu.ent/comm/internal-market as last accessed on 27-7-04

The recommendation applies to all companies covered by EU Accounting Directives allowing for exemptions that member states are permitted to introduce along with these Directives, for small and medium size companies. The recommendation reinforces European Union initiatives in the area of environmental protection. In addition, based on article 15 of the Integrated Pollution Prevention and Control Directive (IPPC), member states will be required to register emission data from large companies (so called IPPC installations) and report this data to the commission. [20] Monitored industrial emissions data should be made publicly available. The application of the International Accounting Standards (IAS) at EU level requires organizations to account for changes to asset values stemming from environmental factor if they are financial (e.g. trading permits).

Federation Des Experts Comptables Europeens (*FEE*) has issued a number of guidelines on EAR including a paper on generally accepted framework for environmental reporting in 1999, detailing objectives, qualitative characteristics and users of environmental information. [21] The *Global Reporting Initiative (GRI)* describes itself as a multi-stakeholder process and independent institution whose mission is to develop and disseminate globally applicable Sustainability Reporting Guidelines. Its Guidelines are for voluntary use by organizations for reporting on the economic, environmental, and social dimensions of their activities, products, and services based on reporting principles. [22]

The *International Standards Organization (ISO)* has developed an extensive range of standards. Among those ISO 14000 series is directly related to the environment. *UN* Global Compact is another initiative that facilitates a network of UN agencies, business, labour, NGOs and governments to promote companies to adhere to ten principles in the areas of human rights, labour, environment, and anti-corruption. [23]

There also exist some guidelines and standards on audit and assurance of environmental statements. The International Standard on Assurance Engagements (ISAE) 3000 is a guideline for audits other than Audits or Reviews of Historical Financial Information developed by the International Auditing and Assurance Standards Board (IAASB) of the International Federation of Accountants (IFAC). ISAE 3000 was first published in December 2003 to be used by accounting firms to guide their assurance engagements on sustainability reports. In Australia, the ISAE 3000 standard has already been accepted for the audit of greenhouse gas emissions under the NSW Greenhouse Gas Abatement Scheme.

In March 2003 the UK-based Accountability issued the AA1000 Assurance Standard (AA1000AS) that covers the full range of an organization's disclosure and performance based on the three core principles of “materiality”, “completeness” and "responsiveness" to ensure that reporting and assurance meets stakeholders' needs and expectations. Some other countries like Australia, Germany, Japan, Sweden etc also have some guidelines/ standards/recommendations on the Verification, Validation and Assurance of Environmental and Sustainability Reports.

In addition, many other national bodies have issued guidelines and awards to encourage environmental reporting and provide guidance on how to establish environmental reporting systems. [24]

3.2 Need for EAR Standards in India

It has been observed in the previous Section that though in most of the countries EAR is still voluntary, governments of some countries have responded to stakeholders' pressure by introducing mandatory disclosure/reporting requirements on environment for companies. Some professional accounting bodies have also issued guidelines on EAR. However, the current financial accounting framework in India appears to be ill-equipped to provide the information required by various internal and external stakeholders on environmental costs and liabilities, environmental risks involved in their investments, social costs or externalities and steps taken by companies to reduce global warming and improve sustainability. This Section examines the need for comprehensive EAR laws in India.

Recent legislation (various Environment Protection laws) in India has expanded the scope of responsibility for pollution prevention and clean up for business organisations. With growing legal and social awareness, there has been a significant increase in environmental costs and liabilities, to the extent that they have become a significant part of total costs and liabilities. Concealing them or their improper disclosure might expose interested stakeholders (like investors, creditors and others) to unknown levels of risk. On the other hand, reporting of this information in annual or environmental reports would enable the concerned stakeholders to get an understanding of the company's stance on environmental conservation, how it specifically deals with environmental issues and in assessing environmental risks involved therein. They can use this information in making relevant investment or other decisions. Hence, it has been increasingly realized that significant environmental costs, benefits, assets and liabilities should be properly measured and reflected in the financial reports.

Since 1991, the concern for environment has increased in India due to economic liberalization and deregulation. After over one and a half decade of reforms, India is entering an era of rapid economic expansion. With the opening of the Indian economy, companies in India are facing strong international competition. These international firms are disclosing non-financial information including environment related facts and figures leading to an enhanced expectation from Indian companies to act responsibly towards the environment and be accountable to the society beyond the traditional role of providing financial account to the shareholders. But the level of environment related disclosures in India, both financial and non-financial, in the corporate annual reports or through other mediums is not yet at encouraging level. Review of literature on the issue has revealed that the corporate financial statements in India lack critical information about environmental impacts or liabilities.

One important reason for this being neither company law nor the accounting standards/guidelines issued by the Institute of Chartered Accountants of India (ICAI) prescribe any accounting or disclosure norms for the environment related issues in the annual or environmental reports. In the absence of clear cut guidelines, business concerns are often uncertain about how to account and incorporate these issues in financial statements. When companies do not provide adequate data to investors, it prevents them from making sound financial decisions. This lack of disclosure also prevents society as a whole from understanding the "global environmental balance sheet," which measures how we are destroying or restoring the support system on which all life on earth depends.

In fact, increasing pollution levels, green house effects, global warming and other related environmental issues have become a matter of increasing international concern these days. Public concerns the world over has resulted in building mechanisms such as the IPCC and the Kyoto Protocol for Greenhouse gas (GHG) reduction. Many countries including India are now considering to control green house effects and CO_2 targets through regulation of business entities and individuals using environmental taxes; carbon emission rationing system where they allocate carbon credits or permits to them for the emission of a certain quantity of greenhouse gases in a particular period (i.e. a permitted quota); or by approving certain organizations as being able to issue legitimate carbon credits (called 'abatement certificates'). The Kyoto Protocol is a 1997 international treaty which came into force in 2005, which has introduced a cap and trade system for the six major greenhouse gasses. The environment policy-related Carbon Trading has entered India with big bang.

Many companies in India have started establishing systems for reducing carbon dioxide emissions in the atmosphere. Many related accounting issues have arisen due to carbon trading which must be answered by accountants in India: one of the most important is the accounting treatment of emission certificates which are intangible assets like patents or copyrights. Valuation of these certificates at cost or at market price, amortization of these certificates, and measurement and recognition of related costs, benefits, assets and liabilities are many questions which are required to be decided by accountants. [25] Unfortunately, in terms of financial reporting, the current financial accounting framework in India is not comprehensive enough to provide the information required by companies to meet the challenge of global warming. This is mainly because accounting information systems based on the accounting equation are not designed to cope with non-monetary measures such as CO_2 sources and sinks. As such, despite the fact that emissions trading has come to India (within and outside the Kyoto Protocol) the accounting standard setters have yet to come up with a standard or guideline to account for such activity.

Regulators in various countries are also taking numerous other steps including fiscal measures to stem the deterioration of ecological environment. These days, environmental taxes as a fiscal policy instrument has also become popular in promoting ecologically sustainable activities in many countries (e.g., eco tax in Germany, climate change levy in many European countries, forestry tax in Brazil and Japan). Environmental taxes can be a way of implementing the polluter pays principle by including environmental costs in the price of goods or services. The idea is that people should pay more for things that are bad environmentally speaking, and pay less for those things which are less harmful to the environment. The system of environmental taxes, known as eco tax system, can complement or avert the need for other regulatory approaches. Unfortunately, India has not paid attention to the system of pollution taxes to deter the polluters. It has been advocated by economists and socialists that Green Economic Principles should be implemented and adhered to in India also. The introduction and likely further development of new environmental taxes is likely to mean that environmental costs become material considerations for company annual reporting; the issues arising from this need to be addressed by the national accounting standard setting bodies.

The above discussion shows that changes in national and international environment have posed many new challenges for accountants in India. However, the accounting profession in India has yet to display significant role in sustainability. Environmental accounting is in preliminary stage in India. The major problem is non-existence of clear cut accounting guidelines and specific accounting standards on the issue. Till date, in India, there is no financial reporting or auditing standard dealing explicitly with social, environmental or sustainability related issues. At present environmental reporting is also not mandatory in India. The only disclosure requirement under company law is regarding conservation of energy measures to be disclosed according to Particulars required under the Companies (Disclosure of Particulars in the Report of Board of Directors) Rules, 1988. It includes disclosure of energy conservation- measures taken, additional investments and proposals for reduction of consumption of energy, impact of these measures and total energy consumption per unit of production. Due to growing social, legal and judicial pressures, some companies have started taking initiatives in this regard, but in the absence of any accounting standard or guideline, identification, measurement and recognition of environmental costs, benefits, assets, and liabilities is, to some extent, judgemental. There is an urgent need, therefore, to adopt a more disciplined and systematic approach towards disclosure of environmental information so as to provide useful, objective and timely environmental information to the concerned stakeholders. In addition, to increase credibility and reliability of environmental information supplied to various interested stakeholders, it is necessary that it should be verified by some independent auditors.

The professional accounting bodies in India should add the issue of EAR to their agenda. An environmental accounting framework is yet to be developed in India which could contribute standards for EAR. The emphasis of this framework should be to raise awareness of environmental issues, to develop guidelines to assist identification, measurement and recognition of environmental issues and evaluation and reporting of these issues, to provide education programs across disciplines focused on environmental issues, and to develop practices of environmental accounting, with recommendations on best practices. Identification and allocation of environmental costs (and cost drivers) and benefits within internal accounting systems (environmental management accounting) should also be considered by companies as well as management accounting institutes.

An effort should be made by the profession to understand and incorporate social, environmental and sustainability related issues into national codes of corporate governance. Professional accounting institutes can also join up with other groups–such as environmental economists–to develop appropriate environmental costing techniques. It is noticeable that in other countries also, many of the advances in environmental accounting, particularly at the management accounting and strategic sustainability accounting levels, are being carried out by non-accounting based organizations, like the United States Environmental Protection Agency (EPA–United States).

The ICAI should also develop a separate standard on Carbon (Emission and Sequestration) Accounting (CES Accounting) which needs to be consistent with the Intergovernmental Panel on Climate Change (IPCC) principles. Other CES measurement and reporting approaches to be considered in this regard are: the Global Reporting Initiative (GRI, 2006); the United Nations Conference on Trade and Development's (UNCTAD) Intergovernmental Working Group of Experts on International Standards of Accounting and Reporting (UNCTAD 2006) and the World Resources Institute and the World Business Council for Sustainable Development (2007) with its Greenhouse Gas Protocol (GHG Protocol). [26] In addition, the measurement of externalities, tax treatment of environmental activities and verification of an entity's environmental accounts by independent duly qualified assurors are some other issues to be considered by the ICAI.

In view of recent developments in the financial marketplace, the ICAI should require an explicit consideration of social and environmental issues (particularly significant environmental cost and liability) at different suitable locations in the annual reports of companies as part of environmental financial accounting/corporate governance. In addition, Indian Companies Act and other related statutes should be amended suitably requiring the companies to disclose information on some other important items like pollution control measures, conservation of natural resources, legal compliance to various Acts, significant environmental liabilities and risks. SEBI may also require listed companies to disclose information on some important items like compliance with environmental laws and pending judicial proceedings arising under environmental laws on prescribed form. For this purpose, a clause in the listing agreement (like clause 49 for Corporate Governance) may be inserted for strict compliance.

In the current scenario, there is also a need to provide formal education and training to the next generation of accountants in the area of EAR. Similarly, there is a need to provide continuing education to already qualified accountants in industry and commerce. It would be best if these issues were fully integrated into the normal accounting, auditing and taxation syllabi of professional accounting bodies. Accountants in India should be well acquainted with the techniques to incorporate environmental issues in annual reports and in methods of auditing sustainability reports. The profession generally could become more obviously involved in management systems certification (e.g. ISO 14001 etc.) and in providing consultancy services in the area of environment or sustainability issues.

For this, a substantial revision to the way in which auditors are educated and trained may be required. . Research into issues such as carbon pricing, life-cycle costing, environmental taxes, sustainability accounting, and carbon efficiency management, would perhaps be avenues by which the accounting profession can play its part in reducing global warming.

3.3 Summary

A significant body of accounting and finance literature has witnessed an increase in the demand for company's environmental information over the past few decades. It has been demonstrated repeatedly that investors now demand disclosure of information regarding emissions (even if legal), failure to comply with environmental regulations and potential liability to environmental remediation requirements. Investors consider such information relevant to their decisions—and thus financially material. However, the availability of information on environmental issues has not kept pace with its growing financial significance. There can be dramatic damages that can be suffered by companies and investors through lack of transparency regarding environmental risks and exposures. If disclosure is not uniform among companies, it will further make investors' job difficult.

The analysis of accounting standards and guidelines in different countries shows that despite a great deal of grassroots activity in the environmental accounting area, the accounting standards setters at the international and national levels have generally chosen to avoid dealing directly with the topic of environmental issues in financial reporting. Though EAR is gradually becoming an important part of corporate accounting and reporting systems, till date, only a few international standards exist concerning how environmental issues may be treated in annual reports. The recommendations of the UN ISAR group and the EU represent worthy attempts by non-accounting organizations to fill the gap and should be taken forward by various accounting standard setters.

The accounting profession in India also has tremendous potential to support industry in developing environmental and sustainability solutions. Both at the individual accountant's level and at the institute level more could be done to manage, account and report on environment related issues. There is an urgent need for development of a standard on EAR to help companies in properly identifying, measuring, recognising and disclosing relevant costs, benefits, assets, liabilities and contingencies relating to environment. The profession should also contribute in this area by developing an environmental and sustainability reporting framework on the lines of those developed by FEE and GRI. A more integrated form of reporting should be developed, in which key environmental, social and sustainability indicators are presented to various internal and external stakeholders for helping them in making informed decisions. However, the basis for all information to be disclosed in the annual report should be to exhibit a true and fair view of the firm's financial situation. Environmental issues must be disclosed in the annual reports if they have or may in the future affect the firm's revenues, costs, or growth or survival of the business. In nutshell, the annual report package as a whole should be enlarged to recognize the new realities of the information demands of the market place.

Notes

1. Pramanik A.K. et al. 'Environmental Accounting and Reporting: With Special Reference to India', *The Cost and Management*, Vol. 35, No.6, (Nov.-Dec 2007), pp. 16-38.
2. Rathore S., *op. cit.*, pp. 418-436.
3. Praminik A.K., *op. cit.*, pp. 119-120.
4. Johnson L.T., "Research on Environmental Reporting", *Accounting Horizons*, 7(3), (Sept. 1993), pp. 118-123.
5. For details on mandatory environmental reporting schemes in 4 selected European countries, see Appendix IV.
6. Canadian Institute of Chartered Accountants, "*Environmental Costs and Liabilities: Accounting and Financial Reporting Issues*", CICA Toronto, (1993), p. 129.
7. Canadian Institute of Chartered Accountants, "*Reporting on Environmental Performance*", CICA, Toronto, (1994).
8. Pramanik A.K., *op. cit.*, p. 18.
9. *Ibid.*, p. 15.
10. The Environment Agency, *Standards and Guidelines*, 2004, Source: http:/environmentagency. gov.uk/business, as last accessed on 12-7-2004.
11. KPMG, '*International Survey on Corporate Responsibility*

Reporting, 2005', Source: http://www.kpmg.nl/Docs/Corporate_site/Publicates/International_Survey_Corporate_Responsibility_2005.pdf. last accessed on 19-7-07.

12. Pramanik A.K., *op. cit.,* p. 19.
13. *Ibid.*
14. Gonzalez C.L., "Steps Towards Mandatory Environmental Accounts in Spain", *Social and Environmental Accounting*, Vol. 19, No. 1, (1999), p. 11.
15. Pramanik A.K., *op. cit.*, p. 25.
16. Environment (Protection) Rules, 1992 of India.
17. Pramanik, A.K., *op. cit.,* p. 29.
18. Abu-Ghazelah T., *op. cit.*
19. EU Recommendation on Measurement and Disclosure on the Recognition of Environmental News in the Annual Accounts and Annual Reports of Companies, May 30, 2001, Source: http://europa.eu.ent/ comm/internal-market as last accessed on 27-7-04
20. KPMG (2002), *op. cit.*
21. FEE (1999), *op. cit.*
22. (www.globalreporting.org).
23. (www.globalcompact.org.)
24. KPMG (2002), *op. cit.*.
25. Ratnatunga, J., "An Inconvenient Truth about Accounting", *Journal of Applied Management Accounting Research, 5*(1), (2007).
26. The GHG protocol is an international accounting tool for government and business leaders to understand, quantify, and manage greenhouse gas emissions. The protocol consists of two modules: *Corporate Accounting and Reporting Standards providing* methodologies to business and other organizations to inventories and report all of the CO_2 emissions they produce and *Project Accounting Protocol and Guidelines,* (*Ratnatunga, 2007*).

4

Existing Studies and Their Critical Evaluation

As mentioned in the previous Chapters, since mid 1990s, there have been significant developments in the area of environmental accounting and reporting (EAR). Professional accounting bodies in many countries are examining the need for a standard or separate conceptual framework on the issue. In many countries (like Denmark, Spain and Australia) disclosure of some environmental information in or outside the financial statements has been made mandatory. However, it is still a voluntary phenomenon in most of the other countries. This area has also attracted the attention of academicians and researchers all over the world. Numerous studies on corporate social reporting and voluntary disclosure practices have been carried out in the past. Environmental issues as a part of corporate social reporting have been addressed in these studies. However, financial implications and accounting challenges associated with environmental issues have been the topic of considerable academic rhetoric and research only since the past decade. After 1990, a number of attempts have been made all over the world, specifically evaluating environmental disclosure practices of companies. But still, only a few research works have addressed the issues relating to environmental accounting and auditing.

This Chapter presents a review of existing literature on EAR in India and abroad. The Chapter is divided into three Sections. Section 4.1 summarises studies on environmental accounting and some related issues. Sections 4.2 reviews the research works done on environmental disclosure practices in different countries. Section 4.3 is concerned with critical evaluation of the existing studies.

4.1 Studies on Environmental Accounting

This Section presents some of the studies on environmental accounting and other related issues conducted in different countries.

Surma and Vondra [1] conducted a survey based on 125 large US corporations in 1990. It was found that companies, which recognised their environmental responsibilities by making board-level committees were viewed more favourably by shareholders. According to the researchers, corporations choosing this strategy would be better positioned in future, when environmental issues would occupy an increasing share of the US regulatory and legislative concerns, public concern and scientific research. They observed that despite a growing concern for the environment, only 14 percent of the responding companies had formal board-level committees. Further, only 11 percent of the companies had accounting policies specifically addressing environmental accounting; less than one-third of them disclosed these policies in the footnotes to the financial statements.

In two of the most comprehensive reports on the subject of pollution prevention in the industrial community, INFORM [2] (a not-for-profit group in USA) studied 29 companies in the organic chemical industry in 1985 and again in 1992. The research found that 'chemical plants with some type of environmental accounting programmes' had an average of three times as many "P2 (pollution prevention) projects" as plants with no cost accounting system. [3] The study also showed that the average annual savings per P2 project in production facilities, where data were available, were just over US$ 3, 51,000, which equalled an average saving of US$ 3.49 for every dollar spent. There were not only substantial savings and returns on investment documented for P2 projects, but an average of 1.6 million pounds of wastes per project were reduced for each project. These results highlighted the potential benefits of environmental accounting to the business community.

Stanko et al. [4] reported a study in 1990, which found that only 11 percent of the companies surveyed had policies that addressed environmental accounting. They found that in the absence of strict accounting pronouncements from the Financial Accounting Standards Board (FASB) and/or the Securities and Exchange Commission (SEC), many firms minimise their financial statement disclosure for environmental loss. A reporting deficiency of this magnitude exposes financial statement users to unknown levels of risks. However, they observed that accounting standard setters were actively evaluating approaches that might improve the reporting of this risk.

In a study to determine the extent to which Chief Executive Officers and Chief Financial Officers used environmental accounting, Wilmhurst and Frost [5] surveyed top 500 Australian companies in 1996. They observed that despite a large number of the respondents indicating that environmental issues were important to their firms and were taken into account in decision-making, in reality, a few companies were actually incorporating environmental issues in a formal way that would allow clear identification and measurement of performance in this area. The survey did not identify the reasons for the lack of formal environmental accounting, but it was suggested that the absence of proper accounting standards and a lack of awareness of the benefits of environmental accounting might be some of the reasons.

A survey administered by Nagle [6] at the Spring 1994 Global Environmental Management Initiative (GEMI) showed that corporate professionals were placing a high priority on environmental accounting. Responses were obtained through questionnaires from 25 respondents. One-half of them stated that their companies had some form of a tracking system for environmental costs. All respondents (except two) reported that they believed environmental accounting issues would be more important to their companies in the near future.

Darryl et al. [7] reported some case studies in corporate

environmental accounting. According to them, these case studies were not about the information that firms reported to the stakeholders, but about the information they gathered for their own purposes and how they used the information. To explore how different firms handled this task, World Resource Institute (WRI) together with teams of academic researchers and corporate staff studied nine companies including Ciba-Geigy and Du Pont. With the help of in-depth interviews and confidential information, the case studies demonstrated how a better accounting of environmental costs led to better business decisions. It is critical because firms look for the ways to produce more from less raw material and energy. When WRI took out environmental costs out of the 'Sundry Accounts' where they had been hidden, the sum was astounding. They found that uncovering these hidden costs could provide significant opportunities for decision-making in production, environmental performance and business planning. They concluded that relying only on conventional accounting, managers might never have known these opportunities.

Collison et al. [8] investigated the integration of environmental performance measurement into core management accounting and financial information systems. The study was based on the results of a questionnaire survey mailed to Finance Directors of Top 200 companies in UK and Top 500 companies in Scotland. A total of 165 replies were received. The broad results were presented by them on aggregative basis. However, some sections of the questionnaire were analyzed by company size and degree of 'environmental sensitivity'. The findings showed an association between the separate identification/quantification of environmental issues and the size and degree of environmental sensitivity of respondents' firms. The survey revealed a variety of viewpoints about the significance of environmental issues for business and a variety of practices in terms of identifying, recording and analyzing the financial implications of environmental factors.

About 27 percent of the respondents cited environmental legal compliance as the main factor affecting their financial position to a great extent. Commercial pressures and voluntary desire were other factors considered significant by them. Only 29 percent of the firms separately quantified environmental costs and benefits. Main reasons for not quantifying these costs and benefits included non-materiality and difficulty in quantification. Some of the firms separately reported the environmental expenditure or revenue mainly through monthly management accounts followed by an environmental report. Majority of the firms considered environmental information relevant in companies' spending decisions (e.g. capital investment appraisal and purchasing of material). The most common accounting treatment of these costs was allocation as overheads. Only a few companies allocated them to product or process. They suggested a need for some accountants to begin to rise to the challenge of the environmental agenda and emphasized the importance of initiatives like that of the UN. They also recommended that a number of companies, particularly small ones, have a scope to improve their awareness and their management accounting practices so as to improve their expertise in highlighting and analyzing environmental issues.

Bebbington et al. [9] (1994) conducted a survey based on a postal questionnaire addressed to finance directors of the 1000 top UK companies (listed on The Times 1000). 350 replies were received of which 181 were usable. The results showed that accountants were aware that environmental issues would affect their practice in the future and perceived that this impact falls within the role of the accountant. They viewed themselves as the appropriate individuals to innovate in this area. Despite this awareness, there was an absence of environmental accounting. They confirmed that there was a low level of environmental accounting activity and accountants were not highly involved in their companies' response to the environmental agenda. They were found to be involved only in traditional accounting areas, such as disclosure. They observed that though the questionnaire respondents worked in companies of different sizes and country of incorporation and country of operation, their attitudes appeared to be homogeneous. Respondents' characteristics (like age and qualifications) also did not affect their responses.

Kokubu et al. [10] examined the influence of two main governmental initiatives on environmental accounting in Japan. These initiatives were taken by the Ministry of Environment (MOE) and the Ministry of Economy, Trade and Industry (METI). They analyzed Japanese corporate environmental accounting practices by an analysis of environmental accounting disclosure and a questionnaire survey. All companies (1,430) listed in the first section of the Tokyo Stock Exchange Market were asked to send a copy of a stand-alone environmental report, and 257 reports published in 2000 were collected. It was found that 184 of these companies (71 percent) disclosed environmental accounting information. The results showed that whereas the publication of environmental reports was not yet widespread in Japan, disclosure of accounting information on environment was quite high among those companies, which published these reports.

The investigation of the information disclosed in these reports indicated the strong influence of MOE guidelines on the classification and definition of environmental costs, which were the main elements of environmental accounting. Then, the questionnaire was sent to those companies, which disclosed environmental information. The decision-making level for environmental accounting, the influence of governmental initiatives, and the purposes and benefits of environmental accounting were examined. It was found that Japanese corporate environmental accounting was oriented to external reporting, since the MOE guidelines stressed the external disclosure function of environmental accounting more strongly than the internal management function. It was concluded that though MOE guidelines strongly influenced Japanese companies and oriented them to external environmental reporting, the METI initiatives for environmental management accounting did not greatly influence them.

Lodhia [11] attempted to explore the vigilance of accountants in Fiji in handling the environmental agenda within their corporate role. The study aimed to ascertain the views of accountants on environmental accounting, assess their preparedness in handling environmental issues within traditional accounting practice, and suggest reasons for their involvement or non involvement in environmental management accounting and reporting in organisations. Opinions were collected with the help of semi-structured interviews conducted at 8 major accounting firms in Fiji including PWC and KPMG. Corporate accountants of both public and private companies were also targeted and the position of their organisations on environmental accounting was established. The findings suggested that Fiji accountants were conspicuously absent from environmental management accounting and reporting in organisations. This was attributed to their lack of competence in environmental matters and the voluntary nature of the present environmental accounting practice.

Thy [12] reported 'The Danish EPA's evaluation of Green Accounts in 1999', which was based on interviews with managers in more than 500 companies, corresponding to over 60 percent of those responsible for the submitted Green Accounts. Furthermore, 550 Green Accounts were analyzed and additional interviews were conduced with 320 selected individuals with a specialized background or a presumed interest in the Green Accounts, and 1900 consumers were also interviewed. It was found that companies were still experimenting with the results that they could achieve from green accounting. The evaluation showed that, in general, external stakeholders demonstrated relatively little interest in obtaining and using Green Accounts. About half the companies which prepared these accounts reported that they gained economic benefits from their work with the Accounts. Companies generally agreed that Green Accounts created a better overview of their own environmental situation. The concerns achieved environmental improvements mainly in energy; water and waste; and consumption of resources. Companies with Green Accounts were found to be better equipped to answer customers' enquiries about documentation on environmental impact and performance. Preparation of these Accounts also enhanced dialogue with employees. The evaluation showed that a large proportion of companies used the obligation to prepare and publish the required information in order also to support internal decision-making. The experience from the first two years showed that both economic and environmental benefits could be achieved through Green Accounts.

In nutshell, a review of studies on environmental accounting reveals that there are only a few reported works on the issue. In India, no study seems to have empirically addressed the issue of environmental accounting. Most of the existing works concentrated mainly on the benefits of adopting an environmental accounting system and checking existence of formal environmental policies and structures. They were quite general in nature and did not focus on specific environmental accounting issues like identification of environmental costs, their capitalisation and amortization. None of the works evaluated the need for a separate conceptual framework or standard on the issue.

4.2 Studies on Environmental Reporting

In this Section, an attempt has been made to review some of the Indian and Overseas studies on corporate environmental disclosure practices. The Section is divided into 3 sub-sections. Sub-section 1 covers some international surveys analysing green reporting practices of companies from more than 2 countries. Sub-section 2 presents some country specific research works on the issue, conducted outside India. Finally, sub-section 3 reviews the research on environmental disclosure practices undertaken in India.

4.2.1 International Studies: Roberts [13] selected companies from five of the largest mainland European countries (namely, Germany, The Netherlands, France, Sweden and Switzerland). 110 large companies were selected who produced full English language version of their annual reports for either 1988 or 1989. It was found that while majority of the companies disclosed at least some environmental information, the extent of disclosure was generally low (with the exception of a few companies who provided extensive disclosures including quantitative measures of outputs and financial costs). No clear country-specific patterns were found in environmental disclosures. The results provided an evidence to support feasibility and practicality of producing such information and suggested that many more companies could provide such information if they so desired.

In 1991, the United Nations intergovernmental working group of experts on international standards of accounting and reporting (ISAR) [14] surveyed 222 transnational corporations to assess the extent to which they were disclosing information on environmental issues. The survey revealed that the disclosure of environmental information by TNCs remained qualitative, descriptive, partial and difficult to compare.

The United Nations [15] undertook another study of environmental disclosures on a sample of 203 annual reports in 1992-1993. Companies were taken from Fortune magazine's 'Global 500' companies representing a number of industries. It was found that 97 percent of the companies disclosed some environmental information. However, vast differences occurred between industries and the types of disclosure made. In general, the chemical, forestry, and petroleum companies disclosed more information and the chemical, and petroleum industries, in particular, usually disclosed their environmental policies.

SustainAbility (the International Institute of Sustainable Development) conducted surveys on corporate environmental reporting in 1993, 1996 and 1997. In the 1993 survey report, SustainAbility [16] summarised interviews with early leading producers of corporate environmental reports in North America, Europe and Japan, with a view to identify trends and developments in the area. The survey found that in North America and Europe, over 85 percent of the companies considered their employees to be the key target audience of their environmental reports. Environmental campaigners were also seen as important in North America and Europe, but less important in Japan. Local communities, shareholders and the media were also mentioned as important user groups. Consumers were considered more important in Europe than in North America. The report gave 2 key conclusions: (i) "decide at the outset which stakeholders you are targeting and find out what they (stakeholders) think and want; (ii) focus your messages and make the format user friendly."

SustainAbility [17] produced another benchmark survey report (this time on behalf of UNEP) in 1996. The report ranked 40 leading companies' corporate environmental reports, out of the (then) estimated 300-400 free standing corporate environmental reports (CERs) available worldwide. 50 reporting ingredients were scored under 5 main categories. After evaluating CERs on this basis, the survey concluded that environmental performance and impact bore little relation to quality of reporting. The three best reports came from very diverse sectors: cosmetics, petroleum and bio-technology.

The next survey report "Engaging Stakeholders" was published by SustainAbility [18] in 1997. The report was a result of a research programme supported by the UNEP and sixteen international companies from Australia, Belgium, Denmark, France, Germany, Sweden, Switzerland, the UK and US. Among the key findings in "Engaging Stakeholders" the following trends were highlighted:

- A new and growing focus on verification, environmental benchmarking, performance indicators, full cost accounting and the implications of sustainability.
- Increase in use of corporate environmental reports (CERs) to monitor, benchmark and rank companies.
- An increasing demand for mandatory – rather than voluntary reporting.
- For leading report-makers, the social dimension of reporting- a critical new area.

Thomas [19] evaluated the extent of voluntary environmental disclosures by multinational corporations (MNCs) during the year 1994. Annual reports of 86 non-U.S. based MNCs were studied to determine the existence, type and extent of environmental disclosures. It was found that though 35 companies in the sample presented some type of environmental disclosure, 42 made no reference to environmental issues. Several types of disclosure were noted, including provisions for environmental liabilities, capital expenditures for environmental upgrades or compliance, and environmental performance goals and goal assessment measures. The disclosures were either positive in tone or neutral in nature. However, no particular trends or patterns were noted by him. In addition, no particular distinguishing characteristics were found between companies that did discuss environmental issues and those that did not.

Thomas and Kenny [20] conducted an exploratory study in 1997 (extending the previous work by Thomas) to investigate and compare the extent and nature of environmental reporting in different forums. In the first phase of the study, they analysed the level of self-reporting of environmental items in the annual reports. For this, they took annual report data from 'Disclosure database' for 86 companies, selected from 500 largest corporations headquartered outside the US. They found that approximately half of the firms in the sample produced no environmental disclosures in the annual reports. Most of the companies reporting voluntary information, made mainly positive disclosures, describing positive steps that the firms had undertaken to be better citizens. They concluded that overall quantity of disclosure was still limited. The second phase of the study was conducted to search the Popular Financial Press to determine the extent of discussion of corporate environmental activities. Taking ABI/INFORM database, they selected 21 cites. It was found that 17 of these cites were positive in nature (congratulating companies on various environmental innovations or planned activities). They observed a paucity of reporting in the financial press, particularly regarding activities that had a negative impact on the environment. They felt that since the study was based on a small sample and relied only on databases for information (not on actual annual reports), most of the conclusions, therefore, could not be generalised.

Scott [21] reported a study in 2000-2001, taking companies from over 30 countries which published environmental and social reports regularly. The study found that though North American companies led the way in the first wave of environmental reporting in the late 1980s due to pressures from regulatory authorities and NGOs, in the 21st century, Europe was the most productive reporting area, with companies from 13 European countries publishing these reports. UK and Germany took the lead in publishing stand-alone environmental reports. Japan got maximum ISO-14001 registrations. While Australian companies were found to be producing many environmental reports, particularly in the mining sector, the performance of other Asian pacific countries was patchy.

Scott [22] reported another study in 2001 revealing the number of environmental, social/community and sustainability reports published by ISO-Technical committee member countries during 1988-2001. It showed that the maximum number of environmental reports were published in UK (504), followed by Sweden (428) and Germany (372). Regarding sustainability reporting, again UK took the lead where 34 companies produced these reports. It was found that many Asian countries (like Bangladesh, Indonesia, Malaysia, Singapore and Sri Lanka) and some others (like Poland, Israel and Turkey) still do not produce any environmental, social or sustainability reports.

KPMG conducted international surveys on environmental reporting at different points of time (1993, 1996, 1999 and 2002). The KPMG (1996) [23] survey showed that environmental reporting was spreading rapidly. The results varied between countries and industries, but the main trend was clear. More and more companies were found to be publishing environmental reports and the quality of the reports was also found improved. The "green glossies" almost disappeared due to stakeholders' demands for quantifiable data and most of the CERs represented a serious commitment by the company to monitor and improve environmental performance.

The results of the KPMG (1999) [24] survey of Environmental Reporting showed that environmental reporting rose significantly in all the European countries surveyed, especially in countries with mandatory reporting rules (e.g., Denmark 29 percent in 1999 as compared to 8 percent in 1996) and a high uptake of EMAS (e.g., Germany 36 percent in 1999 as compared to 28 percent in 1996). The survey discovered that in complete contrast with the European trend, the proportion of top American firms issuing environmental reports fell from 44 percent in 1996 to 30 percent in 1999. Findings of the survey are summarised in Exhibit 4.1.

In a more recent survey by KPMG (2002) [25] Global Sustainability Services, in collaboration with the Graduate Business School of the University of Amsterdam in 2002, environmental disclosure practices of the top companies in 19 countries and the top 250 companies of the Fortune 500 (GFT 250) were analysed. Following similar surveys held before, the 2002 results showed that 45 percent of the GFT 250 companies produced a report on their sustainability or environmental or social performance in their annual financial report. In addition, 28 percent of the top 100 companies in 19 countries issued sustainability reports, compared with 24 percent in 1999. The survey also discovered a shift in focus from merely environmental, health and safety reports to combined environmental, social and full sustainability reports. Among the GFT 250 companies, reporting rates were highest in countries with large corporations, such as the US, Japan, Germany, France and UK.

Exhibit 4.1: Findings of KPMG International Survey of Environmental Reporting, 1999

- A substantial increase in the number of top companies world-wide producing an annual environmental report, from 13 percent of companies in 1993 to 24 percent in 1999, as well as improved quality of reports.
- In contrast to the European trend, the proportion of top American firms issuing environmental reports fell from 44 percent in 1996 to 30 percent in 1999.
- The scope of verification is still varied, and is a long way from a standard which readers can rely on to guarantee the reliability of the reported data and information.
- There is a convergence of administrative and environmental registration systems in major companies, combining the framework and methodology of the financial controllers with the knowledge and expertise of environmental staff.
- Companies are now working towards a balance between financial, environmental and social/ethical performance and are starting to report in all three areas.

Source: Emtairah T., "Corporate Environmental Reporting: Review of Policy Action in Europe", IIIEE Report, Lund, (Feb. 2002), http://www.enviroreporting.com/others/cer-europe.pdf.

Exhibit 4.2 summarises a comparative view of the main findings of KPMG surveys at three different timings.

Exhibit 4.2: Main Findings of KPMG Environmental Reporting Surveys - 1993, 1996, 1999

Findings	1993	1996	1999
No of companies surveyed	810	1300	1100
Response rate (%)	85	69	98
No. with CER	105	220	269
percent of those surveyed	13	15	24
percent externally verified out of those that publish CER	-	15	18
Sustainable development covered (%)	-	12	36
Top 250 companies with CER (%)	-	-	35
Top 250 'polluting' companies with CER (%)	-	-	44
percent of 1100 surveyed companies with environmental information in other corporate reports	-	-	47
Progress with targets (%)	-	26	53

Source: KPMG, 'International Survey on Environmental Reporting', (1999).

However, this reporting trend was also emerging in countries and regions, where it was not practiced earlier (like Latin America). Sectors with relatively high environmental impact, such as chemicals and pharmaceuticals, were leaders in sustainability reporting. However, sectors previously lagging behind (like food and beverage, finance) were catching up on the trend. Of the top 100 companies in each of the 19 countries, Japan had the highest percentage (72 percent) of companies producing corporate sustainability reports, followed by the UK and US. New countries with reporting practices were Hungary, Greece, and South Africa, though the number of reports was still low. In terms of verification, the UK had the highest number of reporting companies with verified reports (26 percent), followed by Japan (19 percent), the Netherlands (10 percent) and Denmark (10 percent).

In its most recent survey on Corporate Responsibility (CR) Reporting 2005, KPMG(2005) [26] analyzed trends in CR reporting of the world's largest corporations, including the top

250 companies of the fortune 500 (Global 250, G 250) and top 100 companies in 16 countries (National 100, N 100). A dramatic change in the type of reporting from purely environmental reporting up to 1999 to sustainability (social, environmental and economic) reporting was found. CR reporting was found to be the mainstream among G 250 companies (68 percent) and fast becoming so among N 100 companies (48 percent). The typical industrial sectors with relatively high environmental impact continued to lead in CR reporting. The percentage of companies providing independent assurance statement also increased as compared to 2002.

Borghini [27] found in a study on environmental information disclosure that all the corporations in Canada and USA disclosed in their annual reports, some kind of environment related information. Italy and the rest of Europe presented high percentages 80 percent and 90 percent respectively, but some European companies (even from polluting sectors such as oil and gas, and chemicals) did not disclose environmental information.

In a study reported by Jones [28] in 2000 (conducted for Commission of the European communities, DG Environment), a survey of Annual/ Financial reports of 88 companies (mainly EU companies or companies that operate in the EU and produce an environmental report) was made. The purpose was to identify the best practice and to evaluate the relevance and comparability of the environmental information that was disclosed. 18 expert organisations and 11 report user organisations that used annual/financial report as a part of their daily business also participated in the survey. The participants were asked to fill a questionnaire. Many report users and expert organisations considered environmental disclosure in these annual reports only a 'green wash' and found that it was difficult to obtain useful and reliable information from these reports. They were of the view that environmental disclosures were difficult to compare across companies as they all disclosed different topics and presented them in different ways. Though environmental disclosures were found in 13 sections identified in the annual reports, the overall level of disclosure by companies was low (except in the environmental section or subsection). 91 percent of the companies disclosed environmental information in the annual report. Though three-fourths of the companies provided this information in the 'environmental section', the space given to the environmental section was quite low (one to four percent of the space in the annual report).

The number of companies disclosing environmental information in the financial report was low. Only one company disclosed information in the income statement, the balance sheet and the director's report. Two-thirds of the companies disclosed nil or limited information in the financial reports. Only a few companies disclosed quantified environmental information and that too in the director's report (e.g., capitalised environmental costs disclosed by 7 companies in the director's report and operational expenses by 2 companies). Some companies disclosed environmental contingent liabilities also.

Kolk et al. [29] undertook a research project to systematically analyze the state of environmental reporting by the 250 largest firms in the world. They analyzed the number and contents of environmental reports. Besides this, they also examined the importance of sectoral differences and firms' nationality. They found that 35 percent of the firms already had an environmental report, while another 32 percent published other types of environmental information. The results indicated that country of origin influenced the frequency of environmental reporting to a great extent. A framework was developed that showed the existence of reporting legislation in firms' home countries (in place/pending/none) against the direct environmental impact of the sector (large or small). Only a small number of reporting firms originated from countries with some kind of legislation (particularly the Netherlands). The national societal pressures seemed to play a large role in these practices (particularly in the UK, The Netherlands and Germany). It was found that there is a clear relationship between direct environmental impact of the sector and the reporting frequency. Non-industrial sectors such as banking and insurance reported less than average. However, there were some exceptions to this impact-reporting linkage, e.g., computer/electronics sector was found to be reporting more on environment. It was concluded that whereas environmental reporting in most of the less-polluting sectors was found to be just starting, the number of externally verified reports increased in those with a substantial direct impact.

4.2.2 Country Specific Studies: This sub-section intends to present reviews of some country-specific studies on environmental disclosure practices conducted abroad.

1. UK: Gray et al. [30] tried to establish relationship between the volume of disclosure and four variables – turnover, capital employed number of employees and profit. Taking social and environmental disclosures in the annual reports of top 100 UK companies (taken from CSEAR data base) for 8 years from 1988-1995 and using regression and longitudinal analysis, they found that in UK, corporate social and environmental disclosure was related to corporate characteristics of size, profitability and industry affiliation. But the study could not find any unique and/or stable relationship between any measure of disclosure and any corporate characteristic. They concluded that in the absence of any unique/stable relationship between amount of variability in disclosure and characteristics given in the study, there was a need for further research to find other factors like organisational culture, experience with pressure groups and media profile.

Harte and Owen [31] found in a survey of UK companies in 1992 that only one company utilised a separate report for disclosing information on social issues. They also found that 40 percent of the companies mentioned social and environmental issues in their statement of corporate objectives and these issues were most commonly found in the chairman's statement or review of activities. Finally, most of the disclosures were narrative in nature and only a few companies referred to 'externally derived standards' when discussing specific areas of performance.

The UK Government's Advisory Committee on Business and Environment (ACBE) [32], in its sixth (1996) progress report, reported that the pace of environmental reporting by UK companies had slackened in the last year or two. Only a few of the largest and almost none of the small companies reported on environment, while the number of companies providing financial information about their environmental impacts in the annual reports remained very small. The study found that "the business case for environmental management is now widely accepted, not only among larger business organisations, but progressively among small and medium sized enterprises as well. Environment reports are aimed to provide information to a diverse readership, but it appears that they are little used by any audience and certainly not by the financial sector because they are not comprehensive and do not provide information for decision-making."

Ince [33] investigated the published statements covering Environmental Policy (EP) of the UK FT–Actuaries 500 companies in 1993/1994. The research took a stakeholders view. A content analysis of the environmental policy (EP) statements of 93 companies was carried out. Results showed that many companies/industries recognised a wide potential range of environmental stakeholders. Companies appeared to pick and choose the areas, which were important to them. The findings also suggested that the industry and size might have affected stakeholder related social and environmental information disclosure in Environmental Policy Statements.

Pension Investment Research Consultant Ltd. (PIRC) [34] in a survey of current practices of FTSE 350 companies (in 1996) found that though 65 percent of the companies reported in some form on environmental issues, there was a wide range of formats, details, and overall approach, even within sectors. The amount of information provided ranged from one sentence to externally verified documents running over 30 pages. Even in sectors with significant environmental impact, such as general industries, reporting was patchy.

Stray and Ballantine [35] selected systematic samples from six industrial sectors (automobiles, banking, electronics, energy, water, and food and drink) in the UK. Questionnaires were sent to 696 companies. Usable replies were received from 277 companies, giving an overall response rate of 33 percent. They found that about one-third of the companies disclosed environmental information and about one-fifth of the remainder planned to do so in the future. Large sized companies were found to disclose more environmental information. 60 percent of the companies did not have environmental policy statement. It was observed that environmental reporting was not confined just to the use of Environmental Reports (ERs) and Annual Reports and Accounts (ARAs). A significant proportion of the companies disclosed this information via other mediums. There was no consensus on the source of guidance to use when providing environmental information either overall or within a specific sector. Substantial differences in environmental disclosures were found between the sectors and moreover, there were considerable variations in reporting practices within individual sectors.

2. Australia: Deegan and Gordon [36] undertook a survey of the environmental reporting practices within the 1991 annual accounts of a random sample of 197 companies from 50 industries. They found that 36 percent of the firms produced environmental disclosures voluntarily. Using word counts, the mean amount of positive environmental disclosures (180.7 words) was found to be significantly more than the mean amount of negative disclosures (5.7 words). Only 14 companies provided any negative environmental information.

Firms in industries which appeared to be environmentally sensitive provided greater positive environmental disclosures. Further, for firms in these industries, there was also a positive correlation between environmental disclosures and firm size. Then, by analyzing annual reports of 25 firms from 16 industries for the years 1980, 1985, 1988 and 1991, they observed a significant increase in environmental disclosures during the period from 1980 to 1991. They linked this change to an apparent increase in societal concern relating to environmental issues. In the third phase of the study, they sent questionnaires to 41 organizations (selected from Green Pages 1991-1992: Director of Environmental Groups in Australia). 22 organizations responded to the questionnaire. From the results of this survey, it appeared that the extent of corporate environmental disclosure was positively associated with the environmental lobby groups' concern about the environmental performance of companies within particular industries.

Deegan and Rankin [37] used a sample of 20 companies, which were successfully prosecuted (for environmental offences) a total of 78 times during the period from 1990 to 1993 and analyzed their annual reports for each of these years. They also reviewed annual reports of another sample of 20 companies, matched by industry and size, which had not been prosecuted, to see if there appeared to be any systematic differences between the environmental disclosure policies of the prosecuted and non-prosecuted companies in the year of prosecution. The amount of environmental disclosure in each case was determined by the use of content analysis, individual words being the unit of measurement. A paired samples t-test and a Wilcoxon matched-pairs signed ranks test were used to determine whether there was any significant difference in the amount of positive and negative environmental disclosures made by the sample firms. It was observed that organizations appeared reluctant to provide any information within their annual reports about any negative environmental implications of their operations. This was the case for both prosecuted and non-prosecuted firms. Those firms that had been prosecuted provided significantly more positive information on environment than their counterparts that had not been prosecuted. This finding supported the view that the prosecuted firms believed that there was a need to counter the negative news of their prosecution with positive news about their environmental initiatives. The existence of a proven environmental offence was reported by only two companies. The results were found to be supportive of the view that environmental information would reasonably be considered as material to account users and hence corporate management emphasized communication of mainly positive environmental information to satisfy the users about environmental performance of the organisation.

Frost [38] reported a study on the environmental reporting practices (ERPs) of 44 Australian companies operating in the extractive industries for the years 1991 and 1994. He investigated the possible casual variables motivating differential levels of environmental reporting. Using content analysis of the corporate annual reports, the ERP of the sample companies were observed. The ERPs were then analysed against the individual entity's public exposure, which was represented by the influence of stakeholders on corporate management and the extent of the entity's interaction with the social environment. Utilising multiple regression, it was found that the public expectation of the entity provided explanatory power at significant levels, for ERP of the sample entities. The results, therefore, suggested that when disturbances such as industry and country of control were held constant, then the level of public exposure was significant in the decision to report environmental information.

In another research study in 2001, Frost [39] investigated the impact of introduction of mandatory corporate environmental reporting in Australia under Sec. 299(1)(f) in 1998. He analysed reporting practices of 71 Australian companies operating in resources (mining, oil and gas), utilities and infrastructure, or paper and packaging industries. Two sets of data were taken, one for the period immediately prior to the introduction of the provision and second for the period immediately following. Results showed that the number of companies disclosing compliance/non-compliance with environmental regulations significantly increased after introduction of Sec. 299(1)(f). More companies disclosed compliance and notably, many more began to recognise non-compliance with the environmental regulations. The results clearly showed that introduction of mandatory environmental reporting significantly improved overall reporting by Australian companies on their environmental performance; contradicting arguments that voluntary and existing disclosure requirements were effective.

In an attempt to explore whether an environmental reporting expectation gap exists within Australia, Deegan and Rankin [40] (in a study sponsored by the Institute of Chartered Accountants in Australia) surveyed the attitudes of senior executives in 462 of the largest companies in Australia (the preparers group) as well as the attitudes of 474 individuals from various categories of annual report users. The main purpose of the study was to explore, whether a potential information demand/ supply imbalance was due to differing perceptions between report users and preparers of these reports as to the relative importance of various items of environmental performance information to the users' decision-making processes. By comparing the responses, significant differences were found to exist between the views of users and the preparers, in relation to various issues associated with corporate environmental performance reporting. The study found evidence showing the existence of an environmental reporting expectation gap within Australia.

Tilt [41] conducted a study on a sample of companies that were listed on the Australian Stock Exchange in 1994, that were in among top 500 Australian companies by market capitalisation and had a formal, documented corporate environmental report (CEP) in place. The CEPs, annual reports and other disclosure items were analysed using content analysis. The study used 'number of sentences' as a measure of disclosure. It was found that annual report was still considered by the companies to be the most appropriate medium for environmental disclosure, demonstrated by their preference for it over other mediums. Other mediums were primarily used to disclose specific activities or particular events of the organisation occurred recently. Even in the annual reports, particular emphasis was put on disclosing activities such as, waste recycling, rehabilitation and pollution control measures; rather than on general policies, research and development, and capital expenditure. When compared to the United Nation's 1994 survey of MNCs, Australian companies appeared to be lagging behind other countries in disclosing their environmental policies.

3. USA: Ingram and Frazier [42] examined annual reports of 40 companies to analyse the association of a company's environmental disclosure with its environmental performance. On the basis of content analysis, they concluded that on an average, the poor performers made slightly more disclosures than those of better performers (except in litigation). Multiple regression analysis also showed that the annual reports of poorer performers contained more environmental information than the better performers. They concluded that the quality of social responsibility disclosure might be a factor affecting utility of data for external monitoring of a firm's social disclosure.

Washington based corporate Rating Organisation-Investment Responsibility Research Centre (IRRC) [43] claimed in early 2000 that US corporations' environmental reporting increased significantly between 1990s and 1999. In their 1999 survey of the Standard and Poor Top 500, IRRC found that 132 of 255 respondents were producing environmental reports.

Similar trends were observed by Gamble et al. [44], who

reported in 1995, an increase in both the quantity and quality of environmental disclosures contained in annual reports and 10-Ks of 234 US corporations. The survey suggested that though there has been some technical guidance in the area from the SEC and the FASB, other less accounting specific factors (like certain environmental accidents, activities by the EPA and other social movements, such as Earth Day) might have provided the major impetus for the increase in environmental reporting.

Al-Tuwaijri et al. [45] provided an integrated analysis of the interrelations among environmental disclosure, environmental performance and economic performance. They used a cross-sectional research design and firm data for the year 1994. The final sample of 198 firms was selected on the basis of predetermined criteria from the 531 firms included in IRRC's 1994 Environmental Profiles Directory. Using simultaneous equations model, they found that "good" environmental performance was significantly associated with "good" economic performance and also with more extensive quantifiable environmental disclosures of specific pollution measures and occurrences. Good environmental performers were found to be disclosing more pollution-related environmental information than the poor performers. Based on the argument that management's (unobservable) overall strategy affects each of these corporate responsibilities, they suggested that prior literature's mixed results describing the inter-relationships between these three variables might be attributable to the fact that researchers did not consider these functions to be jointly determined.

4. Canada: A study undertaken by Stratos Institute [46] (Canada) in 2000 found that some 10 percent of the companies listed on the Toronto Stock Exchange 300 (TSE 300) published detailed environmental, social or sustainability reports. Another 29 companies included two to five pages of environmental and social information within their annual reports, demonstrating some commitment to make this information publicly available. An additional 15 percent included less than two pages of this information in their annual reports. A full two-thirds of the companies still did not include significant environmental or sustainability information in their published reports. In addition, 28 companies not listed on the TSE 300 produced detailed environmental, social or sustainability reports. However, there was considerable variety in what companies reported and how they reported. Almost all the companies presented information on their environmental performance. Most companies reported in a stand-alone published document, while nine integrated this information into their annual reports. Some companies used web-sites for this also. The frequency of report publication also varied from once in two years to once or more than once in a year. The highest proportion of reporters was from electrical utilities sector, followed by chemical companies. Majority of the companies gave voluntary information also.

Buhr and Freedman [47] made a study to compare mandated and voluntary environmental disclosures between Canada and the US, based on document type, nature of information and amount of disclosure. Using 1994 as the disclosure year, 68 Canadian firms from resource based or high environmental impact industries were matched with 68 US firms on a size and industry basis. Annual reports to shareholders, security exchange filings and environmental reports were content analysed and summarised into various categories of information. Using paired t-tests, they concluded that overall, there was no significant difference in the extent of environmental disclosures between US and Canadian companies. Companies in both countries failed to provide adequate environmental disclosures. They noted that US companies provided more mandated disclosure, while Canadian companies provided more voluntary disclosure. It was also found that Canadian companies provided more environmental reports than their US counterparts.

5. Spain: Moneva and Llena [48] analyzed the

environmental reporting practices found in the annual reports published by companies operating in Spain. They also tried to determine the evolution of these practices during the period 1992-94, on the basis of stakeholder theory. The analysis was carried out on the basis of the annual reports of 70 large companies operating in industries considered to be sensitive for the environment. The reporting practices were grouped into a number of categories like type of environmental reporting provided, sections of the annual report devoted to environmental reporting, environmental policies and projects, natural environmental protection activities and achievements, and disclosure of environmental data in the annual accounts. The impact of some variables like sector and foreign association on environmental reporting behaviour was also analyzed. It was found that environmental reporting of the companies was fundamentally narrative in nature, with a very limited use of strictly financial data. However, the presence of quantitative and financial reporting was found to be increasing continuously. No significant differences were found in any of the three years under study with respect to the information disclosed by quoted and non-quoted companies included in the sample. The other factors analyzed were also not found to be significant in affecting reporting practices of the companies. However, in some areas, the sample companies with a foreign parent company disclosed more information during the period of study. They suggested that policy-makers should consider the need to establish environmental disclosure requirements that are useful for the stakeholders.

Larrinaga [49] in a study tried to investigate whether environmental accounting regulation in Spain would be capable of increasing organisations' accountability regarding their environmental impact. His work was concerned with the Spanish standard (1998), which requires all the Spanish companies to include environmental disclosures in their financial statements. For this purpose, a survey of the implementation of environmental disclosure standard was conducted, involving a content analysis of the 1997, 1998 and 1999 financial statements of 70 industrial firms listed on Madrid Stock Exchange. It was found that the standard implementation was deceptive. Approximately 80 percent of the firms surveyed were not disclosing any environmental information at all. The companies that included some environmental information disclosed a mean of 1.8 items, out of a total of seven possible items. Though there was a significant increase in the number of companies disclosing between 1997 and 1999, only 23 percent of the companies disclosed environmental information. Only 2 companies disclosed bad news. The study found that the absence of potentially harmful information in corporate financial statements had to be interpreted as a means to manage public impressions of the firm.

6. Denmark: In a survey of 92 Danish annual reports in 1992-1993, Deloitte Touche [50] found that approximately half of the companies included environmental information, mostly concerning environmental legislation and product development. But in 75 percent of the cases, the environmental information occupied less than 5 lines of the text.

Rikhardsson [51] analysed 156 Danish annual reports in 1993 and reported findings similar to Deloitte Touche. Although 42 corporations included environmental information in their annual reports, this was on a very general level, took up a small amount of space and concerned a limited number of issues. He also attempted to assess the quality of information. He concluded that the information disclosed was generally in qualitative and non-verifiable terms.

Bebbington [52] (1999) found that about 90 percent of Danish firms who were obliged to report under the terms of law, made environmental disclosures. Her evaluation of the Danish experience with mandatory environmental reporting found that reporting appeared to contain mainly positive outcomes. Moreover, a high level of compliance with the law was also found. However, she felt that more work was required in terms of examining quality of reporting.

7. **Malaysia:** Jaffar et al. [53] investigated the environmental reporting practices in annual reports of Malaysian companies. The study enhanced the environmental reporting measurement method used in the previous studies, by using both quality and quantity of reporting. Taking 362 companies listed on First Board of Kuala Lumpur Stock Exchange (KLSE), the study measured environmental information using content analysis. Quality of reporting was measured on the basis of four broad themes or categories namely, environmental policy; product and service; sustainability reporting and audit. Quantity of reporting was measured on the basis of number of pages used to report environmental issues. Multiple regression models were used to analyse the relationship between level of environmental disclosure with environmental performance, size and financial performance. The results showed that environmental information was not well published in the annual reports of Malaysian companies. Most of the information was found in the review of operations and in Chairman's statements. The study also indicated that company's size and environmental performance provided an explanation on the variability of environmental disclosure among companies in Malaysia.

ACCA [54] administered a study in 2002, to know the current status of environmental reporting in Malaysia. The stock of the survey was limited to companies listed on KLSE main board between 1999 and 2001. With the help of a survey of annual reports, stand-alone environmental reports (if any) and Internet research, it was found that there was no statutory requirement in Malaysia requiring public listed companies to disclose environmental information to the public. The survey findings demonstrated that there were an increasing number of listed companies engaging in some form of environmental reporting. The percentage of companies reporting on environment increased from 5.3 percent in 1999 to 7.7 percent in 2001. Half of these companies were from top 100 companies in Malaysia and belonged mainly to industrial product sector. All reporting companies used their published annual reports to communicate this information. Though the quantity of reporting increased overtime; the quality was still limited. None of the reports was verified by a third party. It was concluded that in the absence of any mandatory requirements, environmental reporting in Malaysia was mainly a voluntary initiative.

8. Ireland: O'Dwyer [55] took a sample of 83 listed Irish companies and used content analysis method to ascertain the level of corporate environmental reporting (CER) in the annual report and/or stand-alone publicly available environmental report (if published). He found that 34 percent of the listed companies undertook some form of environmental disclosure. Only one company issued a separate stand-alone environmental report. Disclosing companies tended to concentrate mainly on general environmental policy (18 percent) and products/processes with environmental benefits (12 percent). A number of companies emphasized very broad commitments to environmental protection, but neglected to provide much detail on how commitments were honoured. Only 12 percent of the companies made financial disclosures, mainly to comply with new financial reporting requirements with respect to provisions for environmental liabilities. The results indicated that apart from companies whose core operations had an easily observable environmental impact, there was little evidence of extensive environmental reporting, in terms of either its quantity or quality. The predominance of disclosure incidence among companies in more environmentally sensitive sectors suggested some desire to ensure that they appeared to be concerned about environmental issues. It was also found that while companies did collate information internally regarding their environmental impacts (e.g., for inclusion as a part of an Annual Environmental Report to the Irish Environmental Protection Agency (EPA) in compliance with Integrated Pollution Control (IPC) licensing requirements), they did not engage in any formal external reporting of this information.

Another in-depth study of corporate environmental reporting (CER) practices in Ireland was undertaken by O'Dwyer and Gray [56] in 1998. They conducted a five year longitudinal study (1991-1995) and found that, among the largest Irish listed companies, CER through the corporate annual report lagged significantly behind levels in much of Western Europe. There was also no evidence of the separate production of stand-alone publicly available environmental reports. Over the five-year period examined, the maximum proportion of companies undertaking environmental disclosures in any year was 26 percent. There was an absence of consistency among most of the companies who engaged in CER. For example, a number of companies outlined environmental policies in one year but, failed to do so in subsequent years, thereby, providing no indication as to how policy objectives, which were disclosed, had been achieved (or otherwise). Furthermore, only five companies undertook environmental disclosures in all the five years of the study. They concluded that in the first half of the 1990s, CER in Ireland was very much in its infancy.

9. Other Countries: Niskala and Pretes [57] analyzed changes in corporate environmental reporting practices among large Finnish firms in the year 1987 and 1992. Using content analysis, they analyzed annual reports of 75 Finnish firms drawn from the largest firms in the most environmentally sensitive industries. Each annual report was reviewed to determine whether or not it contained any qualitative, quantitative or financial reporting. The results indicated marked changes in environmental reporting practices between 1987 and 1992. In 1987, slightly over one quarter of the firms analyzed disclosed environmental information in their annual reports, while in 1992 this number had risen to nearly one-half of firms. Most of this disclosure was in qualitative, rather than in quantitative or financial firm. They found results to be consistent with earlier studies, which indicated that environmental reporting was a feature of firms in certain industrial sectors. However, most of the corporations were capable of disclosing environmental information and that a general awakening was found in industry. They saw it as a positive step towards the further development of environmental accounting.

Nikam and Wickramarachchi [58] checked environmental reporting practices in annual reports of Sri Lankan companies. They analysed 24 annual reports of companies out of 240 companies listed on Colombo Stock Exchange for the year 1998-99. They found that annual reports of all the companies in the sample disclosed information on environment, though the number of words used differed. There was a wide disparity among companies in disclosing environmental information. Using descriptive statistics, correlation coefficient, and simple and multiple regressions, it was found that disclosure of physical environmental information was relatively less than social environmental information. It was felt that companies were reluctant to disclose bad news information. Though similar number of companies disclosed good and bad news, the quantity of bad news was quite low. The companies in banks, finance and investment groups had highest level of disclosure; while in hotels and travels had lowest. Positive significant relationship was found between the total value of assets and disclosure of environmental information. Relationship between number of shareholders and environmental information was not significant.

Belal [59] found that in 1996, 90 percent of the selected 30 Bangladeshi companies (of which 28 were listed ones) made some environmental disclosure. Most of the voluntary disclosures were purely descriptive. They were made in notes to the accounts and in the Director's Report. It was found that generally the companies made mandatory disclosures such as expenditure on energy, employees' number and cost.

Imam [60] studied environmental disclosure practices in

Bangladesh from the annual reports of 40 listed companies out of 207 listed companies of the Dhaka Stock Exchange. He found that about 22 percent of the sample companies disclosed environmental information in the annual reports and remaining companies did not show any concern for environmental issues. Environmentally responsible companies also disclosed information only with regard to protection of environment, pollution control, plantation of trees and some other matters. These companies were evidently not aware of issues like waste generation, conservation of energy and water, recycling of waste, odour and noise nuisance. These companies shared only qualitative and positive information. The extent of disclosure remained less than one-fourth of a page or just one or two sentences.

Depoers [61] sent questionnaires to 216 French listed companies' financial directors. The purpose was to examine environmental disclosure decision by taking into account both incentives to disclose and incentives to withhold information. On the basis of 42 replies, he found a strong influence of stakeholders' demand and ethical considerations on disclosure decision. According to the findings, business issues and compliance with what other firms do also encouraged firms to a certain degree, to disclose information. Stock market considerations were found to be relatively unimportant. In addition, results also indicated that the cost/benefit arbitration depended on the nature of the disclosure, e.g., narrative and non-financial items gave rise to the greatest net-benefits. The perceived net costs or net benefits of the items were also found to be significantly different according to the form of communication used, i.e., annual report or environmental report.

Cerin [62] evaluated the practices of the companies listed on the OM Stockholm Exchange to find out the extent to which they produced Documented Environmental Reports on the Internet (DERIs). It was found that 22 firms out of the 88 on the 'A' list and just four out of the 219 listed companies on the 'O' list provided environmental reports on the web. Majority of these companies were large companies. Hence, it was presumed that costs of environmental work were too high for the small and medium enterprises (SMEs). Moreover, it was found that the power, plus chemicals and pharmaceuticals, and forestry sectors had highest proportion of DERI-producing companies on the exchange. However, there was little quantifiable substance in these reports, for instance, in the form of emission performance indicators. It was also found that companies producing DERIs were less eco-efficient than the non-DERI companies. The polluting companies exhibited a greater inclination to report their interest in the environment. The study indicated a risk in taking the decision to invest funds exclusively on the basis of companies' own self-generated external-environmental communication.

4.2.3 Indian Studies: Gautam and Bora [63] studied 25 organisations from Assam and found that 13 of them did not have any concern for environmental information disclosure. 9 companies disclosed descriptive type information, covering space from one-fourth of a page to one page only. 2 organisations presented this information pictorially along with description. Only one company gave some financial information on environment in Social Accounts attached to the annual report. It was concluded that environmental information disclosure by majority of the companies in Assam was confined only to the making of generalised statements with reference to protection of environment, pollution control, conservation of energy, and raw material. They felt that the quality of disclosure was so poor that users could not use it in decision-making.

Kumar [64] examined the Indian Scenario towards environmental disclosures by taking a random sample of 6 giants of Indian corporate sector. As regards the accounting and reporting aspects of environmental protection, sample companies made policy statements in their annual reports. Such information was given in the Chairman's or Director's

Report in the form of statements. Particulars on 'Conservation of energy were given by all the selected companies; but no information was given on expenses incurred, targets set and achieved in respect of natural resources. The study concluded that the environmental reporting in India was still limited to little more than a sentence or two each in the annual report.

Eresi [65] in his study of annual reports of 68 companies for the years 1991-92 and 1992-93 tried to find the number of companies, which disclosed information on environmental matters. The survey showed the disclosure of information on energy conservation by all the companies, as it was a statutory requirement. Among others, the pollution control occupied the first place, followed by protection of environment and raw material conservation. However, the annual reports gave only positive information. Extent of disclosure on environmental issues, except energy, did not exceed even one-fourth of a page in the annual report. It was found that companies generally adopted qualitative (descriptive) form of disclosure except some companies (20 percent), which used pictorial presentation.

Shah et al. [66] analysed reporting on environmental protection activities by public limited companies in India. The study was based on 50 annual reports published by various large and medium size companies in different industries, mainly those which affected environment. It was found that most of the companies formally complied with statutory requirements as to disclosure of energy conservation measures. The companies disclosing on environmental aspects mainly focused on pollution control measures, conservation of energy, effluent treatment, safety arrangements, achievements and awards, and activities like forestation and community health services. Hardly did any company report on failure regarding environmental protection or damage caused by the company to the environment and society. Thus, most of the companies avoided negative reporting. The study concluded that companies had taken reporting on environment more a statutory obligation and less as a social responsibility. The provisions of law on reporting of environmental information are not enough to procure true and fair information about the effect of companies' operations on environment.

Pradhan and Bal [67] collected primary data from a sample of 80 executives during 1997-98, with the help of a structured questionnaire containing 36 questions. The purpose of the study was to examine the perceptions of corporate managers on corporate environmental reporting. They concluded that corporate managers were fully aware of the environmental issues. Majority of them strongly agreed that a company should disclose information on its environmental policy, environmental audit report, quantifiable future goals and targets on environmental issues, disposal of toxic or hazardous substances and on environmental spending. Most of the respondents agreed on the necessity for the reporting of all the 36 items included in the study.

Roy [68] conducted a study to examine the Indian practices in respect of environmental disclosure in corporate annual reports. He took a sample of 55 companies (25 public sector and 30 private sector companies). The annual reports of these companies for three years ending in 1987, 1991 and 1996 were scrutinised. He prepared a checklist of 6 items including pollution control, recycling and reusing disposable material, conservation of energy, development of new energy resources, preservation of ecology of forest, trees and plantation, and other environmental action support programmes. Items of environmental information disclosed in the reports were noted on this checklist. He observed that the common practice followed by the Indian companies, in both the sectors over the years regarding environmental disclosure, was to offer descriptive information and itemize the same in the Director's Report. A few companies disclosed information through notes, schedules and a separate social account. On all environmental issues (except energy conservation), the extent of disclosure did not exceed even one-fourth of a page in most of the cases.

He found that the percentage of companies disclosing environmental information was considerably low in both the sectors. However, the proportion of companies giving this information in the public sectors was more than in private sector.

Vasal [69] administered a study on 144 companies in 1998. He found that 50 percent and 22.2 percent of the companies in the public and private sector respectively disclosed information on their environmental impact through their annual reports. Moreover, the practice of preparing stand-alone environmental reports was yet to take roots in India. Environmental audit was conducted by a very limited number of companies in India.

Thus, a review of existing studies on environmental reporting shows that there are large differences in the environmental disclosure practices of companies in different countries. Several European countries have made systematic effort to increase corporate environmental reporting through specific government requirements and actions. Australia, Spain and Denmark, for example, have all mandated a specified form of environmental reporting for many companies. As a result, environmental reporting in these countries has increased overtime. But it is still in its infancy in many developing countries including India.

4.3 Critical Evaluation of Existing Studies

This Section critically evaluates the existing studies and highlights the salient features of the present work.

While there seems to be no reported study on environmental accounting or auditing in India, the number of such studies even elsewhere has been very small. Some of the existing surveys in different countries (e.g. Surma and Vondra, 1990; Wilmhurst and Frost, 1996; Stanko et al, 1990) examined the existence of environmental accounting systems and policies in the corporate sector. The research by INFORM in 1985 and 1992, mainly concentrated on benefits of being eco-efficient and the importance of environmental accounting in knowing these benefits. Most of the existing attempts on the issue were quite general in nature and did not focus on the specific issues involved in accounting for environmental information (like expensing and capitalisation of environmental costs, depreciation of environmental assets, and recognition and measurement of environmental liabilities).

A review of works on environmental reporting shows that most of them (e.g. Scott, 2001; Guthrie and Parker, 1983; Borghini, 1998) examined the extent to which the companies disclose environmental information in the annual reports. Some studies (e.g. Stratos, 2000; Buhr and Freedman, 1996 and Tilt, 1994) evaluated environmental information disclosure by other mediums like environmental policies, stand-alone corporate environmental reports (CERs) and websites. Some international studies (e.g. by SustainAbility, and KPMG at different points of time) analyzed environmental disclosure practices of the companies from various countries. However, most of the country-specific surveys concentrated on the green reporting practices of the companies in the developed countries. There are only a few attempts from the developing countries (including India). Most of the reported studies in India were made on the basis of a small sample of companies and did not make any detailed statistical analysis of the collected information.

The research works examining the increase in quality of reporting by evaluating the nature of items disclosed, type of disclosure, its length and location in the annual reports are not many. Very few studies (e.g. Gray et al. 1994; Ince 1996; Jaffar et al. 2002) have examined impact of some company and industry related variables on corporate social and environmental disclosures. A preponderance of the research in the area was based on secondary data only and failed to address various conceptual and theoretical considerations involved in reporting on environment. Only a few surveys (e.g. KPMG, 2002; ACCA, 2002) addressed the issue of verification of environmental information reported by the companies. Thus, though there are a large number of studies on environmental reporting, yet only a few surveys have covered environmental accounting and auditing aspects. There is no reported research work covering all the three aspects, namely environmental accounting, reporting, and auditing. In fact, there is a need to cover these issues together as these are interrelated and form an important part of a company's environmental management system. The role of accounting in improving environmental performance of an organisation is also required to be understood.

In subsequent Chapters, the book presents findings of an empirical study conducted by the author on environmental accounting and reporting in India. This exploratory study attempted to overcome most of the limitations of the existing works. The study was global in nature, covering all the three inter-related areas namely, environmental accounting, environmental reporting, and environmental auditing. It was based on primary as well as secondary data. The primary information was collected from a fairly large number of respondents (101 CAs and 101 companies) selected on a predetermined criteria. The study addressed various conceptual and theoretical considerations involved in accounting and reporting of environmental information. The work also examined the environmental reporting practices of the selected companies in the annual reports over three years. The combined impact of selected company- and industry- related variables on environmental disclosure practices was analysed by using multiple regression technique. An attempt was also made to compare findings of the existing study with similar other works, wherever available, conducted in or outside India. It is hoped that the present work would make a significant contribution to the existing knowledge in the area and help in bridging the gap left in the existing available literature.

Notes

1. Surma J.P. and Vondra A.A., "Accounting for Environmental Costs; A Hazardous Subject", *Journal of Accountancy,* (March

1992), pp. 51-55.

2. INFORM, "Cutting Chemical Wastes", (1985); INFORM, "Cutting more Chemical Wastes," (1992), New York, adopted from *EPA, 'An Introduction to Environmental Accounting as a Management Tool – Key Concepts and Terms'*, U.S. EPA primer, Washington DC:EPA, (1995a), p. 28.
3. According to EPA, "Environmental costs often can be reduced or avoided through P2 (Pollution Prevention) Practices such as product design changes, process redesign, input materials substitution, improved operations and maintenance", *Ibid.*
4. Stanko Brian B. and Zeller Thomas L., "Environmental Liability in Financial Reporting", *Business and Economics Review*, Vol. 41, Issue 2 (Jan-March 1995), pp. 19-23, adopted from ABI/INFORM article abstracts on "Environmental Accounting", Source: http:/www.fau.edu/people/students/ paar5428.htm, last accessed on 10-1-05.
5. Wilmhurst T. and Frost R., "Environmental Accounting – a growing concern", *Australian CPA*, (May 1998), Source: http://www.environment.gov.au/ epg/environet/eecp/tools4.html. Last accessed on 28-7-2004.
6. Nagle G., "Business Environmental Cost Accounting Survey," Global Environmental Management Initiative, 1994 Conference proceedings, Arlington,(March 16-17,1994), p. 243.
7. Darryl R., Beth, Beloff and Darryl D., "Case Studies on Corporate Environmental Accounting", Source: http://Green Ledgers in Corporate Environmental Accounting, htm., last accessed on 26-4-03.
8. Collision D. and others, 'Environmental Performance Measurement through Accounting Systems: A Survey of UK Practice', in Bennett M., Rikhardsson P.M. and Schaltegger S., *Environmental Management Accounting – Purpose and Progress*, Kluwer Academic Publishers, The Netherlands (2003), pp. 189-232.
9. Bebbington J., Gray R., Thomson I., Walters D., "Accountants Attitudes and Environmentally –Sensitive Accounting", *Accounting and Business Research*, Vol. 24, Issue 94, (Spring 1994), pp. 109-120.
10. Kokubu K., Nashioka E., Saio K. and Imai S., 'Two Governmental Initiatives on Environmental Management Accounting and Corporate Practices in Japan', in Bennett M. et

al., *op. cit.*, pp. 89-113.

11. Lodhia, S.K. 'Accountants' Responses to the Environmental Agenda in a Developing Nation: An Initial Study on Fiji' (2000), Source: http://ecocomm.anu.edu.au/bim/staff/lodhia/lodhia.html, last accessed on 25-7-04.
12. Thy C., 'The Danish Green Accounts: Experiences and Internal Effects', in Bennett et al., *op. cit.*, pp. 115-127.
13. Roberts C.B., 'Environmental Disclosures: A Note on Reporting Practices in Mainland Europe', *Accounting, Auditing & Accountability Journal*, Vol. 4, No. 3 (1991), pp. 62-71.
14. United Nations, '*Environmental Accounting : Current Issues, Abstracts and Bibliography*', United Nations, New York , (1992a).
15. United Nations, '*Environmental Disclosures: International Survey of Corporate Reporting Practices*', Report of the UNCTAD Secretariat, (E/C.A.s10/A(3)1994/4), New York, (12 Jan.1994).
16. DTTI, IISD, & Sustain Ability, "Coming Clean – Corporate Environmental Reporting, Opening up for Sustainable Development", (1993), DTTI: London.
17. SustainAbility & UNEP, (1996), adopted from Wheeler and Elkington, '*The Recent History of Environmental and Social Reporting*', Source: http:/www.sustainability.com./programs/engaging/history-reporting. asp, last accessed on 29-1-02.
18. SustainAbility & UNEP, "*Engaging Stakeholders : the 1997 Benchmark Survey*", The Third International Progress Report on Company Environmental Reporting, London, (1997), SustainAbility Ltd., Source: www.sustainability.com.
19. Thomas P.B., 'An Exploratory Study of the Extent of Environmental Disclosures by Multinational Corporations', (1996), Source: http://www3.bus.osaka-cu.ac.jp/cpa96/papers.htm/thomas. htm. last accessed on 15-7-04.
20. Thomas P.B. and Kenney S.Y., 'Environmental Reporting: A Comparison of Annual Report Disclosures and Popular Financial Press Commentary', (1997), Source: http://les.man.ac.uk/ipa97/ papers/thomas61.html, last accessed on 13-12-04.
21. Scott P., "Reporting All Over the World", *Environmental*

Finance, (Dec.2000 – Jan. 2001), pp. 36-37.

22. Scott P., "ISO Backs Environmental Reporting", *Environmental Finance*, (Nov. 2001), pp. 20-21.
23. KPMG, '*KPMG International Survey of Environmental Reporting, 1996*', Source: www.wimm.hl/ publication/kpmg 1999.pdf, last accessed on 20-9-04.
24. KPMG, '*KPMG International Survey on Environmental Reporting, 1999*', Source: *Ibid.*
25. KPMG, '*KPMG International Survey of Corporate Sustainability Reporting 2002'*, (2002), Source: http://www.wimm.nl/publicates/KPMG2002.pdf. Last accessed on 10-2-05.
26. KPMG. '*International Survey on Corporate Responsibility Reporting*, 2005'. Source: http://www.kpmg.nl/Docs/Corporate_site/ Publicates/International_Survey_Corporate_Responsibility_2005.pdf. Last accessed on 19-7-07.
27. Borghini S., 'Environmental Information in Annual Reports', *Environmental Communication Observatory*, No. 2, (Sep. 1998), Milane.
28. Jones K., '*Study on Environmental Reporting by Companies'*, Commission of European Communities, DG Environment, Centre for Environmental Informatics, University of Sunderland, (March 2000).
29. Kolk A., Walhain S. and Wateringen S.V., "Environmental Reporting by the Fortune Global 250: Exploring the Influence of Nationality and Sector", *Business Strategy and the Environment*, 10, (2001), pp. 15-28.
30. Gray R., Jarad M., Power D., and Sinclair C.D., 'Social and Environmental Disclosure and Corporate Characteristics: A Research Note and Extension', *Journal of Business Finance and Accounting*, 28(3) & (4), (April/May 2001), pp. 327-356.
31. Harte G. and Owen D., 'Current Trends in the Reporting of Green Issues in the Annual Reports of United Kingdom Companies', in Owen D., *Green Reporting- Accountancy and the Challenges of the Nineties,* Chapter 8, Chapman and Hall, London, UK, (1992), pp.166-200.
32. ACBE (1996), Adopted from a report by Doyale, "*Environmental Accounting Not Widely Used,*" Source: www.ecobody.com/views10.html, last accessed on 12-3-04.

33. Ince D., "Determinants of Social and Environmental Disclosures in the UK Companies in Environmental Policy Statements", Source: http://les.man.ac.uk/IPA97/papers/ince119.pdf.last accessed on 12-3-04.
34. Pensions and Investment Research Consultants (PIRC), 'Environmental Reporting 1996: The PIRC Survey of FTSE 350', PIRC, (1996).
35. Stray S. and Ballantine J., "A Sectoral Comparison of Corporate Environmental Reporting and Disclosure", *Eco-Management and Auditing*, 7, (Dec. 2000), pp. 165-177.
36. Deegan C. and Gordon B., "A Study of Environmental Disclosure Practices of Australian Corporations", *Accounting and Business Research*, Vol. 26, No. 3 (1996), pp. 187-199.
37. Deegan C. and Rankin M., "Do Australian Companies Report Environmental News Objectivity?", *Accounting, Auditing & Accountability Journal*, Vol. 9, No. 2 (1996), pp. 50-67.
38. Frost G.R., "Environmental Reporting in Australian Extractive Industry", (1998), Source: http://www.bham.ac.uk/EAA/eaa98/abstracts/sea/frost.htm.
39. Frost G.R., "An Investigation of the Introduction of Mandatory Environmental Reporting in Australia", 3rd Asia Pacific Interdisciplinary Research in Accounting (APIRA) Conference, (15-17 July 2001), Adelaide.
40. Deegan C. and Rankin M., "The Environmental Reporting Expectations Gap: Australian Evidence", *British Accounting Review*, Vol. 31(3), (1999), pp.313-346.
41. Tilt A., "The Content and Disclosure of Australian Corporate Environmental Policies", Source: www.ssh.flinders.edu.au/commerce/researchpapers/aa.4.htm., last accessed on 30-1 2005.
42. Ingram R. and Frazier K., "Environmental Performances and Corporate Disclosure", *Journal of Accounting Research*, Autumn, (1980), pp. 614-622.
43. Investment Responsibility Research Centre (IRRC), (2000), adopted from Wheeler and Elkington, *op. cit.*
44. Gamble G.O., Hsu K., Kite D. and Radtke R., "Environmental Disclosures in Annual Reports and 10 Ks: An Examination", *Accounting Horizons*, 9(3), (1995), pp.34-54.
45. Al-Tuwaijri S.A., Christensen T.E. and Hughes K.E., "The

Relations among Environmental Disclosure, Environmental Performance, and Economic Performance: A Simultaneous Equations Approach", *Accounting, Organizations and Society*, 29, (2004), pp. 447-471.

46. Stratos, '*A Snapshot of Sustainability Reporting in Canada*', Canada, (2001), Source: www.stratos-sts.com/sts-files/stratos.chapter.2.pdf., last accessed on 15-1-2005.
47. Buhr and Freedman, "A Comparison of Mandated and Voluntary Environmental Disclosure", Source: http://panoptic.csustan.edu/cpa 96/txt/buhr.txt,198., last accessed on 17-8-04
48. Moneva J.M. and Llena F., "Environmental Disclosures in the Annual Reports of Large Companies in Spain", *The European Accounting Review*, 9:1 (2000), pp. 7-29.
49. Larrinaga C., "An Evaluation of Mandatory Environmental Accounts in Spain", Source: visar.csustan.edu/papers/larrinage 91.pdf. , Last accessed on 30-1 05.
50. Deloitte Touche, "Environmental Reporting in Annual Reports", Deloitle Touche Tohmatsu International Copenhagen, Danish Environmental Ministry (1995), Source: Green Reporting. Danish Environment Min.
51. Rikardsson P.M., "The Measurement and Reporting of Corporate Environmental Information", A Paper prepared for the Nordic Network Conference, 'Business and the Environment, Sandvika, 7-12 Dec. 1994'.
52. Bebbington J., "Compulsory Environmental Reporting in Denmark: An Evaluation", *Social and Environmental Accounting*, Vol. 19(2), (1999), pp. 2-4.
53. Jaffar R., Iskandar T.M. and Muhamad M., "An investigation of Environmental Disclosure in Malaysia", Source: www.aaanz.org/web2002/accepted0.020-papers/jafar.pdf, as last accessed on 30-1-05.
54. ACCA, "Mandatory Environmental Accounting in Malaysia", Source: www.acca.co.uk/pdfs/ environment/other/TECH-ERM-001-5.pdf, Last accessed on 30-1-05.
55. O' Dwyer B., "The Ponderous Evolution of Corporate Environmental Reporting in Ireland: Recent Evidence from Publicly Listed Companies", *Corporate Social Responsibility and Environmental Management*, 10 (2003), pp. 91-100.
56. O'Dwyer B. and Gray R.H., "Corporate Social Reporting in the

Republic of Ireland: A Longitudinal Study", *The Irish Accounting Review*, 5(2), (1998), pp. 1-34.

57. Niskala M. and Pretes M., "Environmental Reporting in Finland: A Note on the Use of Annual Reports", *Accounting, Organizations and Society*, Vol. 20, No. 5 (1995), pp. 457-466.
58. Nikam G.A. and Wickramarachchi M.W., 'Environmental Reporting and its Application in Listed Companies in Sri Lanka', in Pramanik A.K., *op. cit.*, pp. 406-423.
59. Belal A.R., "Corporate Social Reporting in Bangladesh", *Social and Environmental Accounting*, Vol.19 (1), (1999), pp.8-12.
60. Imam A., "Environmental Reporting in Bangladesh", *Social and Environmental Accounting*, Vol. 19(2), (1999), pp.12-14.
61. Depoers F., "An Empirical Investigation of Environmental Disclosure Strategy of French Listed Companies", a paper presented at Euro conference project, Financial Reporting and Recognition Practices in Europe (FIRRE), (May 20-23,2001), Altavilla Milicia, (Sicily, Italy).
62. Cerin P., "Characteristics of Environmental Reporters on the OM Stockholm Exchange", *Business Strategy and the Environment*, 11 (2002), pp. 298-311.
63. Gautam H.C. and Bora B., 'Environmental Accounting and its Significance in the Context of Corporate Information Disclosure', in Pramanik A.K., *op. cit.*, pp. 320-333.
64. Kumar P., 'Environmental Accounting: An Indian Panorama', *Ibid*, pp. 75-79.
65. Eresi K., "Information Disclosure in Annual Reports," *The Chartered Accountant*, (Jan.1996), pp. 45-48.
66. Shah G.S., Patel C.P. and Shah F.H., ' Reporting on Environmental Aspects by Joint Stock Companies- A Sample Study', in Pramanik A.K., *op. cit.*, pp. 140-144.
67. Pradhan B.B. and Bal R.K., 'Corporate Environmental Reporting; Perceptions of Corporate Managers', *Ibid,* pp. 313-320.
68. Roy M.N., 'Environmental Disclosure in the Corporate Annual Reports', in Rao P.S.B. and Rao P.M., *Environment Management and Audit,* Deep and Deep Publications Pvt. Ltd., Delhi, (2000), pp. 228-244.
69. Vasal V., "Accounting for Sustainable Economic Development: Issues and Evidences", *Indian Journal of Accounting,* Vol. XXXI, (June 1998), pp. 12-34.

5

Corporate Environmental Accounting and Reporting Expectation Gap: Evidence from India

It is evident from the previous Chapters that accounting of environmental matters has been emerging as an important dimension of corporate accounting and reporting practices. In many countries, the disclosure of environmental information has been made mandatory. Companies, all over the world, have also started reporting on the matter voluntarily. In the wake of growing legal and social pressures, many companies in India have also made attempts to disclose their concern for the environment in the annual reports or by some other mediums like environmental statements or Internet. But, in the absence of any specific law, accounting standard or guideline on the issue, these disclosures are subjective, inconsistent, and mainly descriptive and positive in nature. In fact, recent research on corporate environmental reporting has shown that environmental reporting in India is typically deficient and not of a standard to satisfy the information needs of various classes of report users. [1]

This Chapter explores whether a potential demand/supply imbalance is due to differing perceptions between report users and preparers as to the various issues associated with Environmental accounting and reporting (EAR) and importance of environmental information to the users' decision-making processes. For this purpose, the study surveyed the attitudes of senior executives in 101 large manufacturing companies in India (the preparers group) on various issues relating to EAR. An attempt was also made to obtain information on EAR practices of these companies from the executives. In addition, opinions of the selected individual Chartered Accountants (CAs), who were annual report users in different capacities, were obtained. An attempt was made to examine whether an EAR expectation gap exists within India. The Chapter is structured as follows. In Section 5.1, scope and methodology used is given. In Section 5.2, general concept of environment has been discussed. In Section 5.3, opinions of the CAs and the executives on various issues relating to environmental accounting have been presented. Section 5.4 presents a comparative analysis of the views of the CAs on environmental information disclosure by companies and the executives' perceptions of their companies' environmental reporting practices. Section 5.5 deals with audit of environmental information reported by concerns in the annual reports or environmental statements. Finally, Section 5.6 summarizes the Chapter.

5.1 Scope and Methodology

This exploratory study dealing with EAR in India is based on primary data. One of the objectives of the study was to examine the need for a specific regulatory framework (including accounting guidelines, principles and standards) for EAR in India. For this purpose, two questionnaires were designed keeping in mind guidelines issued by professional accounting bodies like the United Nations' Intergovernmental Working Group of Experts on International Standards of Accounting and Reporting (ISAR), Financial Accounting Standards Board (FASB), International Accounting Standards Committee (IASC), Canadian Institute of Chartered Accountants (CICA) and Federation des Experts Comptables Europeans (FEE). The questionnaires covered some general and specific issues relating to environmental accounting, environmental audit and disclosure of environmental information by companies.

The first questionnaire was used to obtain opinions of chartered accountants (CAs) on various issues relating to EAR.

200 CAs were selected from a directory prepared by the Institute of Chartered Accountants of India (ICAI) and the questionnaires were mailed to them. Only two percent of the questionnaires were returned. Then, the CAs in Delhi, NCR region and some major cities were personally contacted. 118 responses were received, of which 101 usable ones were used in final analysis. The profile of the responding CAs showed that they were mature, highly experienced and were related to the corporate world in different capacities. They had sufficient knowledge of the corporate accounting and reporting practices (as judged from their engagement in company audits) and were capable of rendering opinions on various aspects of EAR due to their diverse knowledge and experience.

To analyze EAR practices followed by companies, another structured questionnaire was framed on the same lines. The questionnaire was pilot tested in 10 large companies having registered/corporate offices situated in Delhi. It was found that none of these companies had a practice of preparing environmental accounts for external reporting purposes, though some of them used environmental cost information for internal decision-making. It was also observed that most of them disclosed some environmental information (though mainly qualitative) in the annual reports generally through the Director's Report. In view of these major findings, the questionnaire was adjusted to cover opinion seeking questions on environmental accounting in addition to questions on actual environmental disclosure practices. For rest of the survey, only this questionnaire was used. An attempt was also made to obtain opinions of the executives on various issues involved.

The companies for the study were selected from the Delhi Stock Exchange's list of companies in India ranked on the basis of their annual turnover. Top 500 companies were taken from this list. Manufacturing companies from both public and private sectors were approached for the survey. Out of these 500 companies, 161 non-manufacturing concerns (other than those operating in power generation and mining industries) [2] were excluded as the issue of environmental accounting might not be of much relevance to them; thus, giving a sample of 339 companies for the study. Since response received through mail was not encouraging, personal visits were made to the corporate/registered offices of companies situated in Delhi and adjoining areas, Mumbai, Kolkatta and Chennai. The response received was overwhelming. 105 responses were received, of which four were rejected due to non-availability of the required information. Thus, 101 questionnaires were used for the final analysis, giving a response rate of 29.8 percent.

An idea about the representative nature of the study can be obtained by looking at the characteristics of the companies. The reliability of the information supplied can be judged by scanning the experience, designation, age and academic qualifications of the respondents. The maximum percentage of companies (about 17 percent) were from 'general engineering and automobiles' industry, followed by 'iron & steel and other metals' and 'chemical, fertilizers and pharmaceuticals' industries. A significant number of the companies (11) were from 'oil, gas and petrochemicals' industry, which is a polluting industry. In fact, an attempt was made in the study to include more companies from the industries which are considered to be the 'most polluting industries' according to the Environment (Protection) Rules, 1992, as the issue of EAR is more relevant for them. In total, 58.42 percent of the companies were from the most polluting industries. There were 20 public sector and 81 private sector companies in the sample. Exhibit 5.1 shows industry as well as sector-wise classification of the companies.

The Exhibit indicates that companies from both the sectors were there in the sample, depending upon the nature of industry and degree of Government or Private Player's participation in that industry. Thus, the study fairly represents various industries and sectors. The responding companies were also found to be representative of the corporate sector in India by looking at their other characteristics (e.g., region, paid-up capital and foreign association).

Exhibit 5.1: Industry and Sector-wise Classification of the Companies

Number (Percent)

Industries	Sector	
	Public	**Private**
Coal and Mining, Cement	2 (33.3)	4 (66.7)
Textiles	-	5 (100)
Paper and Paper Products	2 (66.7)	1 (33.3)
Electronics, Electric Equipments	-	9 (100)
Iron & Steel and other metals	3 (20)	12 (80)
General Engineering, and Automobiles	1 (5.9)	16 (94.1)
Chemicals, Fertilizers and Pharmaceuticals	2 (13.3)	13 (86.7)
Sugar, Breweries, Food Products and FMCG	-	5 (100)
Oil & Gas and Petrochemicals	7 (63.6)	4 (36.4)
Power Generation	3 (75)	1 (25)
Others (Diversified, Paints, tyres etc.)	-	11 (100)
Total	**20 (19.8)**	**81 (80.2)**

Note: The percentage of companies in public and private sectors in each of the industries is given within parentheses.

Regarding qualifications of the responding executives, about 80 percent of the managers had professional degrees (like CA, CS, CWA etc.) in commerce. Therefore, relevant information from well qualified and knowledgeable executives was obtained. The questionnaires were filled by mature and highly experienced senior officials. About 42 percent of the executives were from the '40-50 years' age group and 27 executives were having age above 50 years. About one-half of the executives had a total experience of more than 20 years in the industry. Majority of the respondents held strategic positions (like Director, President, VP or GM of

Accounts/Finance Department) in the companies. The profile increases the reliability of the information provided for the study.

5.2 Concept of Environment

In literature, the term environment has been elucidated by various authors differently. In this study, the term has been defined as natural physical surroundings and includes air, water, land, flora, fauna and non-renewable resources such as fossil fuels and minerals. In recent years, the environmental crisis has become a global issue. Now-a-days, there is a greater degree of awareness and concern among people about environmental degradation. This is evident from the fact that besides governmental agencies, a large number of non-governmental organisations (NGOs) are also involved in the process of environmental protection at global, national, regional and local levels. This Section seeks to present opinions of the CAs and the executives on some general issues relating to environment. The Section is further divided into two sub-sections. Sub-section 1 records responses of the CAs and the managers on the importance of environment as a challenge for business. Sub-section 2 deals with need for introducing environmental management system (EMS) in an organisation.

5.2.1 Environment - A Big Challenge to Business: In large number of major environmental problems like deforestation, over-exploitation of resources and pollution, industrial and business activities are directly or indirectly responsible. Because of growing awareness, business and industry cannot remain a silent spectator to all these things. They have to take care of environment to survive and grow. Majority of the CAs strongly agreed that environment is one of the most important contemporary challenges faced by business. In their opinion, a business takes basic resources from the environment and grows and survives in it. In turn, if it does not take steps to protect the environment, it cannot continue to operate in the long run. This shows that senior accounting professionals were aware of the increasing importance of environmental issues. Only three CAs were of the opinion that the main objective of a business is profit-making and issues relating to profit generation are more important for a business. A business has to solve many other challenges relating to finance, competition and customers. Environment comes only after solving these problems.

An overwhelming majority (about 97 percent) of the executives also agreed that environment is one of the most important challenges faced by business these days. About seven-tenths of them strongly believed this. They were of the opinion that due to growing public awareness, governmental and non-governmental organizations as well as judicial bodies have become sensitive towards environmental issues. Moreover, if a company has to operate in international market, it needs to prove that it is a green company and therefore, environment is one of the crucial issues in these companies.

5.2.2 Establishment of EMS: Business and industry are also increasingly realising their role in sustainable development and hence, a new area of management is fast emerging known as 'Corporate Environmental Management'. [3] Environmental impact assessment, environmental risk management, environmental accounting and environmental audit are some of the important areas falling within the scope of environmental management. The opinions of the CAs on the need for introducing EMS in a business were obtained in the study. Almost all the CAs (99 percent) agreed that companies should establish EMS. Majority of them (57.5 percent) strongly recommended this. The CAs felt that establishment of a good EMS helps in maintaining clean and green environment around the factories, controlling pollution emission and thereby meeting legal standards in this area. It facilitates compliance with various legal requirements set by Pollution Control Boards (PCBs). Establishment of EMS is also a pre-condition to obtain ISO-14001 certificate, which most of the big companies are getting these days. Some of them felt that EMS helps in satisfying the government, the NGOs and the society at large that business is meeting its social responsibilities. Only one respondent felt that establishment of EMS is a costly affair and businesses, particularly small entities, do not require it.

An examination of actual practices showed that a large majority of the companies (89.11 percent) already had EMS in their organizations, while another four companies were planning to introduce it in near future. The executives gave various reasons for their companies' decisions to establish EMS like 'to meet legal requirements relating to environment', 'concern for environment', 'required for obtaining ISO-14000 certificates', 'growing legal, social and even financial pressures'. The executives belonging to the polluting industries felt that existence of an efficient system of environmental management helps in satisfying regulatory authorities about environmental concern of the organization. Only seven companies didn't have a formal EMS in place. The main reason advanced for not introducing EMS was that the company belonged to a non-polluting industry.

5.3 Environmental Accounting

The previous Section has revealed that a large majority of the respondents considered environment as one of the most important challenges faced by business these days. Increasing social, legal, global and financial pressures have led these companies to initiate steps to protect and conserve environment. These measures involve some costs, but at the same time, give many direct or indirect benefits to the company as well as to the society. To provide for impact of these steps on trading results and financial position of a company, it is necessary to introduce some form of accounting for the environment. It has been observed in Chapter 2 that broadly the concept of environmental accounting can be discussed at two levels- National Accounting level (macro level of environmental accounting) and Corporate Accounting level (micro level of environmental accounting). Corporate level again includes 2 parts–Environmental financial accounting (used for external reporting purposes) and Environmental management accounting (used for internal decision-making). The present study covers environmental accounting at corporate level only, i.e. financial accounting level and management accounting level. Environmental accounting has been defined as identification, measurement, recognition and disclosure of environmental costs, benefits, assets and liabilities.

In this Section, an attempt has been made to present environmental accounting practices of the selected companies. The opinions of the CAs and the senior preparers of accounts were also obtained on some specific issues relating to accounting for the environment. This Section is further divided into 8 sub-sections. Sub-section 1 seeks to find whether existing financial accounting framework is comprehensive enough to deal with environmental issues. Sub-section 2 analyzes environmental accounting expectation gap in India by comparing opinions of the CAs regarding adoption of environmental accounting system by companies with actual environmental accounting practices of the sample companies. Sub-section 3 presents opinions of the respondents on the importance of environmental information in decision-making. Sub-section 4 deals with identification of environmental costs. Sub-section 5 records viewpoints of the respondents on treatment of these costs in the books of accounts. This Sub-section also takes up some specific issues relating to environmental accounting like treatment of prior-period environmental costs and recovery and impairment of environmental assets. Sub-section 6 covers the recognition of environmental benefits. Sub-section 7 discusses treatment of social costs by the companies. Finally, Sub-section 8 analyses opinions of the respondents on recognition and measurement of environmental liabilities.

5.3.1 Comprehensiveness of Financial Accounting Framework: Conventionally financial statements are prepared within the framework provided by recognized standard setting bodies in various countries. These statements are prepared for investors, creditors and other stakeholders. However, the incorporation of environmental costs, benefits and other issues into the mainstream of financial accounting is still at an embryonic stage. It is commonly advocated that the existing accounting framework is not comprehensive enough to deal with environmental costs, benefits, assets, liabilities and contingencies. Till date, in India, neither Company Law nor any accounting standard prescribes any accounting treatment or disclosure technique for environmental matters in the corporate financial statements. This sub-section attempts to present opinions of the senior preparers of accounts on the adequacy of the existing financial accounting framework to deal with environmental issues.

More than four-fifths of the executives felt that the current financial accounting framework in India can not deal with all the issues relating to accounting for the environment. 88 percent of the CAs were also of the same opinion. These accountants supported development of a separate framework for environmental accounting to find ways to monitor the use and value of environmental resources both in terms of raw material consumed and damage inflicted upon environment. In their opinion, though the financial accounting framework can be used broadly for dealing with general environmental issues, some specific environment related problems (like separating environmental costs from other costs or measurement of some less tangible environmental costs and benefits) cannot be handled effectively within this framework. For handling these problems, a separate environmental accounting framework is required within the broad financial accounting framework.

Further about 83 percent of the executives and 88.6 percent of the CAs supported development of a separate standard on environmental accounting or amendments in the existing standards to incorporate environmental matters. They felt that this would help in dealing with specific issues relating to accounting for environment. They were of the opinion that in the absence of specific standards, these issues are handled in the books of accounts according to the discretion of the managers and it may lead to manipulation of results. However, some of the respondents were of the opinion that only a few adjustments in the existing financial accounting framework can make it adequate enough to deal with environmental problems.

5.3.2 Adoption of Environmental Accounting System: The existing research on the issue shows that till date, very few companies incorporate environmental accounts formally in their financial statements. Though some companies have started reporting on environmental issues, the preparation of environmental accounts for financial reporting purposes is still not common. This sub-section records opinions of the CAs on the main reasons for failure of many companies to incorporate environmental accounting formally in their financial statements. Further, actual environmental accounting practices of the responding companies have also been examined in this part.

According to the CAs, the most important reason for failure to prepare environmental accounts is 'voluntary nature of environmental accounting'. About 63 percent of the qualified accountants gave one of the first three ranks to this reason. Two more reasons considered important and given first three ranks by about 55 percent of the CAs were 'ignorance of benefits of environmental accounting' and 'lack of clear-cut government policy on this issue'. The CAs were of the view that companies in India tend to lack initiative in this matter. They would introduce environmental accounting only if it is made mandatory. Moreover, people are still not aware of the benefits of environmental accounting (Exhibit 5.2).

In the opinion of the respondents, clear-cut government policy and framing of a separate accounting standard on the issue would be essential to prompt companies to use this concept. Some respondents gave other reasons also like, 'costs involved in preparation of environmental accounts', 'reluctance on part of top-management', 'insensitivity towards environmental issues' and 'no direct benefit to companies'.

Exhibit 5.2: Reasons for Failure to Use Environmental Accounting: CAs' Views

Number (percent)

Reasons	Ranks						
	0	1	2	3	4	5	6
(a) Lack of accounting standards	20 (19.8)	14 (13.9)	22 (21.8)	18 (17.8)	11 (10.9)	14 (13.9)	2 (2)
(b) Ignorance of benefits of environmental accounting	14 (13.8)	24 (23.8)	10 (9.9)	22 (21.8)	19 (18.8)	11 (10.9)	-
(c) Voluntary nature of environmental accounting	15 (14.9)	34 (33.7)	14 (13.9)	16 (15.8)	9 (8.9)	13 (12.9)	-
(d) Lack of clear cut government's policy on the issue	19 (18.8)	23 (22.8)	18 (17.8)	15 (14.9)	19 (18.8)	7 (6.9)	-
(e) Difficulty in measuring environmental costs and benefits objectively	17 (16.8)	17 (16.8)	19 (18.8)	12 (11.9)	15 (14.9)	21 (20.8)	-
(f) Any other	89 (88.1)	11 (10.9)	-	-	-	-	1 (1)

The CAs was further asked whether a company should adopt environmental accounting system and if yes, what are the potential benefits of adopting this system. Almost all the CAs (97.03 percent) felt that a company should use environmental accounting system. One of the senior practicing CA was of the opinion, "Though at this stage, it seems difficult for a company to use this system, in the long run, preparation of environmental accounts is beneficial not only for the concern, but also for various stakeholders who are interested in environmental information".

Majority of the qualified accountants considered 'helps in complying with environmental laws' as an important benefit of preparing environmental accounts (Exhibit 5.3). The other two benefits regarded important by them were, 'helps management in decision-making' and 'provides better estimates of cost of producing a product'. Some CAs mentioned other benefits of preparing these accounts also like 'it helps business in meeting its social responsibilities', 'required in global business perspective' and 'helps in enhancing image of the company'.

Only three CAs didn't find any need for preparing separate environmental accounts mainly because it is not required by law. In their opinion, preparation of environmental accounts is costly as well as a time-consuming affair and is not suitable for entities particularly small business. Moreover, in the absence of any accounting standard or guideline, it is difficult to measure environmental information. In such a scenario, preparation of environmental accounts will not serve any purpose.

Actual Practices: The practices of the sample companies relating to accounting for the environment have been depicted in Exhibit 5.4.

The Exhibit shows that a substantially high proportion of the companies (93.1 percent) did not prepare environment related accounts. About three-fourths of the managers gave non-mandatory nature of environmental accounting as a justification for not preparing these accounts (Exhibit 5.5).

Exhibit 5.3: Benefits of Environmental Accounting: CAs' Opinions

Number (Percent)

Reasons	Ranks			
	0	1	2	3
(a) Provides better estimates of total cost of producing a product	22 (22.7)	16 (16.5)	22 (22.7)	15 (15.5)
(b) Helps management in decision-making	17 (17.5)	26 (26.8)	18 (18.6)	13 (13.4)
(c) Improves pricing of the product	35 (36.1)	2 (2.1)	8 (8.3)	19 (19.6)
(d) Increases profitability	36 (37.1)	3 (3.1)	4 (4.1)	5 (5.2)
(e) Gives competitive advantage	22 (22.7)	9 (9.3)	12 (12.4)	16 (16.5)
(f) Helps in complying with environmental laws	9 (9.3)	41 (42.3)	19 (19.6)	7 (7.2)
(g) Any other	86 (88.7)	10 (10.3)	-	-

Contd...

Reasons	Ranks		
	4	5	6
(a) Provides better estimates of total cost of producing a product	9 (9.3)	10 (10.3)	3 (3.1)
(b) Helps management in decision-making	14 (14.4	7 (7.2)	2 (2.1)
(c) Improves pricing of the product	11 (11.3)	17 (17.50	5 (5.2)
(d) Increases profitability	13 (13.4)	10 (10.3)	25 (22.8)
(e) Gives competitive advantage	14 (14.4)	17 (17.5)	7 (7.2)
(f) Helps in complying with environmental laws	5 (5.2)	2 (2.1)	14 (14.4)
(g) Any other	-	-	1 (1.0)

Exhibit 5.4: Preparation of Environmental Accounts by Sample Companies

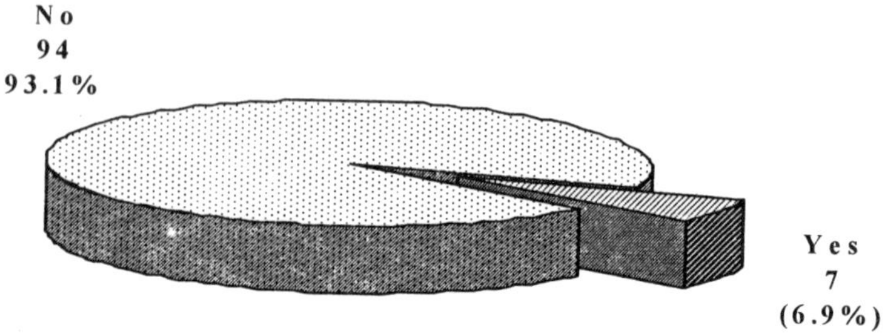

About 24 percent of the executives mentioned 'no accounting standard and guideline' as the main reason for not preparing these accounts. In the absence of any specific standard or guideline in this area, these officials found it difficult to incorporate environmental issues formally in the annual accounts. In this situation, if environmental statements are prepared, there will be lack of any credibility, objectivity and comparability in the information provided in these statements. It is to be noted that a negligible proportion of the managers advanced cost and time constraints as explanations for not preparing these statements. Some executives gave other reasons also like 'inadequate database', 'no effect on profitability', 'costing problems'.

Only seven executives informed that their companies prepared environmental accounts. But during interviews and discussions, it was found that even these companies did not have practice of incorporating environmental accounts in the financial statements for external reporting purposes. However, these companies generated some environmental figures for internal decision-making purposes. Further discussions on the issue revealed that about 57 percent of the companies had practice of generating and using internal environmental statements for managerial decision-making.

Exhibit 5.5: Reasons for not preparing Environmental Accounts: Executives' Perceptions

Number (Percent)

Reasons	Ranks					
	0	1	2	3	4	5
(a) Expensive	69 (72.6)	2 (2.1)	-	3 (3.2)	4 (4.2)	17 (17.9)
(b) Time consuming	71 (74.7)	2 (2.1)	1 (1.1)	3 (3.2)	14 (14.7)	4 (4.2)
(c) Not required by Law	5 (5.3)	72 (75.8)	12 (12.6)	4 (4.2)	1 (1.1)	1 (1.1)
(d) No accounting standard or guideline	43 (45.3)	23 (24.2)	21 (22.1)	3 (3.2)	3 (3.2)	2 (2.2)
(e) Difficulty in measuring environmental costs and benefits	61 (64.2)	14 (14.7)	7 (7.4)	12 (12.6)	1 (1.1)	-
(f) Any other	76 (83.5)	10 (11)	4 (4.4)	1 (1.1)	-	-

Thus, a significant environmental accounting expectation gap exists in India. Though about 97 percent of the CAs felt that environmental accounts should be prepared, but 93 percent of the companies did not prepare these accounts for external reporting purposes.

5.3.3 Environmental Costs: Different interpretations of the term environmental costs have been given in Chapter 2 (refer to Section 2.2.3). Some professional bodies and authors include external as well as internal costs in the definition of environmental costs. However, the present study is concerned with internal costs only, which a company actually incurs on protection or conservation of environment (whether voluntarily or under legal requirement). It, therefore, ignores social/external environmental costs. In the present study, environmental costs have been defined as costs of steps taken or required to be taken by a company, to reduce adverse impact of its activities on the environment as well as costs driven by the environmental objectives of the enterprise.

In this sub-section, an attempt has been made to examine views of the respondents on the importance of environmental cost information in making various business decisions. A list of five vital business decisions was given in the questionnaire. The respondents were asked whether environmental information generated by environmental accounting will be useful for management in making these decisions. Their views have been presented in Exhibits 5.6 and 5.7.

Exhibit 5.6: Importance of Environmental Cost Information in Decision-making: CAs' Views

Number (Percent)

Business Decisions	Yes	No	N.A
(a) Appraisal of investment for environmental risks	90 (89.11)	8 (7.92)	3 (2.97)
(b) Evaluation of environmental performance of a company	88 (87.13)	5 (4.95)	8 (7.92)
(c) Planning cost reductions	59 (58.42)	35 (34.65)	7 (6.93)
(d) Assessing environmental impact of a company's projects	91 (90.1)	4 (3.96)	6 (5.94)
(e) Designing various processes using environmental friendly technologies	97 (96.04)	1 (0.99)	3 (2.97)
(f) Any other	-	-	-

The above Exhibit shows that an overwhelming majority (96.04 percent) of the CAs found usefulness of environmental cost information in 'designing various processes using environmental friendly technologies'. According to EPA also, "The design of a process or product significantly affects environmental costs and performance. Many companies are adopting 'design for the environment' or 'life cycle design' programme to take environmental considerations into accounts at an early stage. Thus, making environmental cost and performance information available to designers can facilitate the design of environmentally preferable processes and products." [4] About 90 percent of the CAs felt that this information facilitates making other decisions also, e.g., environmental impact assessment or appraisal of investments for environmental risks. Only 59 responding CAs recognised usefulness of these details in 'planning cost reductions'. [5]

Exhibit 5.7: Importance of Environmental Cost Information in Decision-making: Executives' Opinions

Number (Percent)

Decision Areas	Yes	No	N.A.
(a) Appraisal of investments for environmental risks	93 (92.08)	3 (2.97)	5 (4.95)
(b) Evaluation of environmental performance of a company	94 (93.07)	3 (2.97)	4 (3.96)
(c) Planning cost reductions	71 (70.30)	25 (24.75)	5 (4.95)
(d) Assessing environmental impact of company's projects	98 (97.03)	3 (2.97)	-
(e) Designing various processes using environmental friendly technologies	94 (93.07)	3 (2.97)	4 (3.96)
(f) Any other	-	-	-

An analysis of the responses given in Exhibit 5.7 shows that all but three executives were convinced about the usefulness of environmental cost information in 'assessing environmental impact of a company's projects'. Other important decisions where these details can be fruitfully used according to them were, 'appraisal of investment for environmental risks', 'evaluation of environmental performance of a company' and 'designing of various processes using environmental friendly technologies'.

5.3.4 Identification of Environmental Costs: One of the major barriers to the adoption of cleaner production technologies and eco-efficiency is that firms do not often know the full environmental costs of operating their business. As a result, they do not know the financial benefits that can arise by reducing their environmental impacts (by using cleaner technologies, environmental friendly processes or waste management). The identification of environmental expenses not only provides a comprehensive picture of the efforts that have been undertaken by a business to protect the environment, but also helps in its proper treatment in books of accounts. Some expenses like those incurred on compliance with environmental laws or purchase of pollution control equipment can easily be identified as environmental costs. But others cannot be easily categorized as environmental costs like cost of purchasing equipment which is energy efficient. These costs are normal business costs, but some additional amount may have to be spent keeping in mind environmental considerations. It is difficult to segregate these costs into normal business expenditure and environmental costs. The views of the respondents on the need to bifurcate these expenses into ordinary and environmental costs were collected in the study.

About 88 percent of the responding CAs and two-thirds of the officials felt that these combined costs should be bifurcated into environmental and normal business expenses. In their opinion, segregation of combined cost is necessary for getting information about total environmental costs incurred by a company. If a business really wants to know true environmental costs and report them to the stakeholders, it needs to separate all environmental costs from other business expenses. [6] Some of them found it beneficial in assessing environmental performance of a company and in fixing responsibilities of different managers. The executives in environmentally sensitive industries felt that this segregation gives an idea about a company's total spending on protection and improvement of environment and this information is required to reduce pressures from various governmental and non-governmental bodies. Some respondents were not in favour of bifurcation of combined costs as it is not required by law. They also felt that in the absence of any accounting standard or guideline, this segregation may lead to confusion and manipulation of accounts.

Majority of those respondents who favoured bifurcation of environmental and non-environmental expenses considered 'some accounting standard or guideline' as the most appropriate basis for bifurcation. In their opinion, without a standard or guideline, the bifurcation will give discretion to management and may lead to manipulation of accounts. [7] 'Internal mechanism' for segregation of combined expenses was preferred by some of the executives. They were of the view that companies should define what should constitute an environmental cost and how to classify it, on the basis of their goals and intended uses of environmental accounting. For example, if a company wants to encourage pollution prevention in capital budgeting, it might consider distinguishing environmental costs that can be avoided by pollution prevention investments from environmental costs related to remedying environmental contamination that has already occurred. But for product costing purposes, such a distinction might not be necessary because both are costs of producing the goods or service. Four executives suggested that a combination of above methods can be used according to the circumstances of the case.

5.3.5 Treatment of Environmental Costs in Books of Accounts: After identifying environmental costs, the next step is their proper recognition in the books of accounts. This sub-section presents the views of the respondents on treatment of these costs in the books of accounts, i.e. expensing, capitalisation and amortization of environmental expenses. Exhibit 5.8 summarizes these opinions.

Exhibit 5.8: Treatment of Environmental Costs in Financial Statements

Number (Percent)*

Treatment in the Financial Statements	Responses (Yes)	
	CAs	Executives
1. Expensing of environmental expenses with no future benefits	93 (92.1)	94 (93.07)
2. Capitalisation of Environmental Costs incurred on		
(a) Increasing capacity of an existing asset	91 (90.1)	88 (87.13)
(b) Improving safety of an existing asset	61 (60.4)	53 (52.48)
(c) Improving efficiency of an existing asset	75 (74.3)	73 (72.28)
(d) Reducing environmental contamination likely to occur as a result of future operations	60 (59.4)	66 (65.35)
(e) Preventing environmental pollution	60 (59.4)	64 (63.37)
(f) Conserving the environment	62 (61.4)	62 (61.39)
3. Including related environmental costs in the cost of an existing asset	74 (73.3)	60 (59.4)
4. Amortisation of environmental costs		
(a) 1-2 years	8 (9.1)	4 (4.4)
(b) 3-4 years	30 (34.4)	19 (18.8)
(c) any other frequency	-	-
(d) (d) as per standards	49 (56.3)	69 (76.6)
5. Treatment of pollution permits as intangible assets	43 (42.6)	28 (27.7)
6. Treatment of environmental costs relating to prior period as per existing accounting standards	81 (80.2)	86 (85.1)

* The number and percentage in the Exhibit shows the respondents answering in affirmative to a given treatment of expenses.

1. Expensing of Environmental Costs: As per financial accounting principles, the expenses which are recurring in nature and do not give future economic benefits to the concern should be expensed in the year in which they are incurred.

According to ISAR guidelines also, an environmental cost that does not meet the criteria for recognition as an asset should be charged to Profit and Loss Account. [8] As shown in the Exhibit, majority of the respondents felt that the same financial accounting principle should be followed in case of environmental costs also. Only some managers argued that generally environmental layouts involve huge amount and if charged to one year's revenue account, it will reduce the profits of that year substantially. Therefore, these costs should be treated as deferred revenue expenditure. Moreover, non-recurring environmental costs can be capitalized, even when they do not give future benefits to the concern. [9]

2. Capitalization of Environmental Costs: According to ISAR guidelines, "Environmental costs should be capitalised, if they relate, directly or indirectly, to future economic benefits that will flow to an enterprise through:

- Increasing the capacity or improving the safety or efficiency of other assets owned by the enterprise;
- Reducing or preventing environmental contamination likely to occur as a result of future operations; or
- Conserving the environment." [10]

The Exhibit shows that a large majority of the CAs and the executives were of the view that environmental expenditures incurred on 'increasing capacity of an existing asset' should be capitalised. These costs lead to future economic benefits for a company and hence, meet the criteria for recognition as an asset. Most of them also favoured capitalization of environmental costs for 'improving efficiency of an existing asset' as it will increase future economic benefits either through cost reduction or increase in revenues. [11] Regarding all other environmental costs, only about 60 percent of the respondents favoured their capitalisation. According to them these costs (e.g., cost incurred on conserving the environment or improving safety of an existing asset) though, do not lead to future economic benefits, should be capitalised as they result in benefits to society in future. [12] Some of them argued that increasing legal and social pressures have compelled the companies to incur these costs. Since these costs are necessary for the very existence of the company, these should be capitalised. They were of the view that if these expenses involve substantial outlays and result in asset creation, these should be capitalized even when company is not getting future economic benefits (for example, expenditure in establishing Effluent Treatment Plants (ETPs)).

About 40 percent of the respondents did not favour capitalisation of these expenses. They felt that the only criterion for recognition of an asset is that it should result in flow of economic benefits to the concern in future. They were of the opinion that even if some asset is created, it is a kind of NPA (non-performing asset) and hence, need not be shown as an asset in the Balance Sheet. Moreover, charging these expenses to the revenue account will give tax benefits to the company.

In nutshell, any expenditure for survival and continuation of business is necessary for the very existence of the business and hence, should be capitalised because it would give indirect long term benefits to the concern. Without incurring these costs, the business can not be continued in the future. Therefore, keeping in mind going concern assumption, these expenses should be capitalised. But routine or recurring expenses should be charged to the revenue account in the same year in which they are incurred.

3. Depreciation/Amortization of Environmental Assets: The capitalised environmental costs (i.e., environmental assets) should be amortized over a reasonable period of time. Majority of the respondents favoured amortization over a short period of time. They were of the view that due to rapidly changing environmental laws and technologies for environmental protection, it is financially prudent to write-off these assets as early as possible. The responding CAs and the managers emphasised the need for specific guidelines or standard about the period and method of amortization of environmental assets. These guidelines may be given either in the Company Law (like other assets) or in relevant accounting standards.

4. Treatment of Pollution Permits as Intangible Assets: Many countries are now considering to control green house effects and CO_2 targets through regulation of business entities using carbon emission rationing system where they allocate carbon credits or permits to them for the emission of a certain quantity of greenhouse gases in a particular period (i.e. a permitted quota), or by approving certain organizations as being able to issue legitimate carbon credits (called 'abatement certificates'). The environment policy-related carbon trading [13] has entered India also with many companies in India have started establishing systems for reducing carbon dioxide emissions in the atmosphere. Many related accounting issues have arisen due to carbon trading which must be answered by accountants in India- one of the most important is the accounting treatment of emission certificates which are intangible assets like patents or copyrights.

The respondents were asked whether pollution permits or emission rights should be treated as intangible assets. About 55 percent of the responding accountants and seven- tenths of the executives did not support treatment of pollution permits or emission rights as intangible assets. The executives informed that generally acquiring pollution permits or emission rights is a routine thing for factories. Manufacturing concerns have to get these permits to comply with various environmental laws. Therefore, expenses incurred in getting these permits should be treated as recurring cost and charged to P&L Account in the same year only.

Even those respondents who favoured treatment of these permits and rights as intangible assets said that this should be done only when these rights are acquired for a longer period. Most of the managers strongly recommended that government should give these rights for a longer period because acquiring them every year is a hassle and wastage of time and resources.

They were of the view that these permits should at least be given for three years, though yearly environmental audit by recognized authorities can be conducted to see that pollution or emission level is within the prescribed limits. In case of default, the permit can be cancelled by the concerned State Pollution Control Board (SCPB).

5. Prior Period Costs: In many cases, an environmental cost may relate to the damage that has occurred in a prior period, e.g., an activity in a prior period which now requires clean up or cost of disposing or treating hazardous waste created in a prior period. Current accounting standards specify that a cost related to a prior period should not be treated as prior period adjustment unless there is some fundamental accounting error. [14] This part presents the views of the responding CAs and the executives on treatment of these costs in the books of accounts.

The respondents generally favoured application of current accounting standards here i.e., environmental costs that relate to prior period/periods should not be treated as prior period adjustment unless there is a fundamental accounting error. However, some respondents did not support treatment of the expenses related to a prior period as current expenses. There are many reasons for their viewpoint. Firstly, in their opinion, some companies pay a significant portion of their total environmental cost to clean up pollution caused decades ago. These expenses are often substantial and including them in current year's profit statement will reduce profits drastically. Secondly, loading these expenses into product cost fails to accurately measure the profitability of the product, facility or division. The respondents felt that managers should not be held responsible for the costs beyond their control. Some executives were of the view that if these costs are significant and benefits relating to them have been derived in the past, these costs can be treated as prior period costs.

6. Recovery and Impairment of Environmental Assets: According to ISAR guidelines, environmental assets should be tested for impairment and should be written down to their recoverable amount. Similarly, when an environmental cost is capitalised and included as an integral part of another asset, the combined asset should be tested for impairment and, wherever appropriate, written down to its recoverable amount. [15] This part presents opinions of the accountants on recovery and impairment of environmental assets.

The respondents generally favoured testing of environmental assets for impairment. However, some of them believed that while recognition and measurement of environmental asset impairment involves the same principles as other forms of impairment, the uncertainties may be greater here. About 75 percent of them felt that if appropriate, environmental assets should be written down to their recoverable amount. Almost identical responses were received in case of recovery and impairment of combined assets. It is to be noted that about one-tenth of the CAs was indecisive on the issue. Some of the respondents, who favoured testing of environmental assets for impairment, did not support writing these assets in the financial statements at their recoverable values. They felt that it is against the going-concern assumption. However, some of them said that the recoverable values can be shown in the footnotes.

5.3.6 Recognition of Environmental Benefits: Most of the environmental costs incurred by a company do not result in direct and immediate financial benefits to the company, although they may give benefits to the society. But, there are some environmental costs that give financial benefits to the company in future, e.g., use of energy efficient equipment and machines may lead to lesser running cost. Similarly, steps taken to conserve natural resources or for managing waste may give financial benefits to the company. [16] In a given period, there can be three types of environmental benefits: income, current savings and cost avoidance. [17] It has been observed that companies generally talk about continuously increasing environmental costs, but even financial benefits resulting from them are never recognised in the financial statements or otherwise. Eco-efficiency suggests that in addition to reporting environmental costs, it is also necessary to report environmental benefits, so that the user of data can get full information to take relevant decisions. By comparing benefits produced with environmental costs incurred in a given period, a type of environmental financial statement can be produced. Managers can use this statement to assess environmental performance of the company and potential for progress. This sub-section records opinions of the CAs and the executives on the recognition of expected financial benefits of environmental expenses (e.g., potential cost savings from using cleaner technologies) in the financial statements.

64.4 percent of the CAs did not support recognition of future economic benefits of environmental expenses. Most of them felt that due to the 'principle of conservatism', expected future benefits from environmental costs should be ignored. Other reasons given by some of them were, 'no accounting standard requires it', 'no need to recognise it separately in financial statements', 'financial benefits not to be considered in incurring social costs' and 'intangible nature of most of the environmental benefits. [18]

More than one-third of the CAs was of the view that current environmental financial benefits should be recognised in the financial statements. They gave different reasons for their viewpoint like, 'it will help in cost-benefit analysis', 'helps in assessing environmental performance of a concern' and 'will encourage use of environmental accounting concept'. These accountants were in favour of disclosing current financial benefits from environmental costs in the financial statements and future financial benefits either by way of footnotes to the financial statements or in separate environmental statements. [19] In their opinion, this information is very useful for external stakeholders as well as managers in decision-making. The viewpoints of the CAs on the issue have been summarized in Exhibit 5.9.

Exhibit 5.9: Reasons for Recognition of Environmental Benefits: CAs' Views

Reasons	Number of CAs
A. For recognizing environmental benefits	
(a) Helps in cost-benefit analysis	8
(b) Helps in assessing company's environmental performance	3
(c) Required under principle of full disclosure	1
(d) Any other	6
B. For not recognizing environmental benefits	
(a) Principle of conservatism	42
(b) Automatic accounting in financial statements	4
(c) Financial benefits not to be considered in incurring social costs	4
(d) No accounting standard requires in it	2

About 53 percent of the executives were of the view that these benefits should be separately recognized in the financial statements. They gave various reasons for their viewpoint like 'it is necessary for cost-benefit analysis' and 'will motivatc a company to spend more on environment'. However, about 46 percent of the respondents did not support recognition of these benefits in the financial statements because of differing reasons like 'difficulty in measurement'; 'lack of accounting standards' and 'principle of conservatism'. The viewpoints of these executives have been summarized in Exhibit 5.10.

5.3.7 Social/External Costs: Social costs are costs of adverse impact of a business's activities on the society, for which it is not legally accountable (e.g., environmental degradation or adverse impact on the health of human beings, for which business is not legally liable). Recognition and measurement of these costs is a difficult thing as these costs are qualitative in nature and also, no accounting standard or guideline provides any recommendation in this regard. Though methods of valuation/measurement of these costs are not in the scope of this study, an attempt has been made in this part to analyse general opinions of the respondents on the issue. [20] Their views have been presented in Exhibit 5.11.

Exhibit 5.10: Reasons for Recognition of Environmental Benefits: Executives' Opinion

Reasons	Number of Executives
A. For recognizing environmental benefits	
(a) Highlights activities of the company for environment	4
(b) Essential for the concept of environmental accounting	2
(c) Necessary for cost-benefit analysis	9
(d) Will motivate company to spend more on environment	6
(e) Will provide environment related information to stakeholders	15
B. For not recognizing environmental benefits	
(a) Difficulty in measuring	11
(b) Lack of accounting standards	4
(c) Automatic recording is there	5
(d) External disclosure not necessary	10
(e) Principle of conservatism	10

As depicted in the Exhibit, a large majority of the CAs (96 percent) felt that a company should try to determine external adverse impacts of its activities. They were of the opinion that business is a part of society and as a good corporate citizen, it is a duty of the company to see that its activities are not harmful for the society. One senior qualified accountant was of the view, 'Such an accounting will give human face to the company and will add to its competitive position in the market.' Only four CAs were not in favour of determining social costs because of difficulties involved in it. They also argued that determination of these costs is not required by law.

Exhibit 5.11: Social Costs/Externalities

Number (Percent)

Area	Response (Yes)	
	CAs' views	Actual practices
1. Determination of Social Costs	97 (96.1)	81 (80.2)
2. Valuation of Social Costs	88 (87.1)	42 (41.6)
3. Considering Social Costs in Decision-making	94 (93.1)	90 (89.1)
4. Internalising social costs	96 (95.05)	97 (96.1)

Regarding actual practices of the company, about 80 percent of the companies determined external adverse impact of their activities. Only about 17 percent of the companies informed that they did not follow the practice of determining these costs mainly because of less polluting nature of the company. Some of the companies did not follow this practice as it was not required by law.

The CAs were further asked whether a company should try to value these social costs. As evident from the Exhibit, majority (87.1 percent) of the CAs felt that though it is difficult to value external costs, some method should be used by companies for this valuation/measurement. In their opinion, only after valuation of these costs, a company can think of reducing these costs or compensating the society for adverse impacts of its activities. Only nine CAs felt that companies should not waste time and resources in valuation of external costs because, firstly, there is no accounting standard on valuation of social cost and secondly, law does not require it. An overwhelming majority of the CAs (93 percent) favoured consideration of social/external costs in making various strategic business decisions like before starting a new project or using a technology.

When actual practices were analysed, it was found that more than 53 percent of the companies did not try to value these costs. The executives belonging to these companies argued that in the absence of any accounting standard or measuring technique, it was not only difficult, but useless to value these social costs. However, 42 executives informed that their companies tried to evaluate these externalities and accordingly took steps to mitigate adversity of these impacts. They believed that valuation of social costs helped their companies to decide about the amount to be spent on these issues. [21] A large majority of the companies claimed that social costs are considered in making crucial business-decisions. The senior executives informed that considering social costs is sometimes a legal compulsion, but many a times, this is done voluntarily also keeping in mind social and environmental objectives of the company. Environmental impact assessment is an important part of project planning and management.

The boundaries of social costs are not static. Many costs, which were external/ social in the past, are now internal costs for a business due to changes in legislation. In the same way, due to growing legal and social pressures, most of the external environmental costs may become internal in future (e.g., companies will have to bear penalties or clean up pollutions). Almost all the qualified accountants and the senior executives were of the view that most of these external costs will become internal in future. They felt that society is becoming aware of the environmental issues and unless a business itself takes some voluntary measures to prevent and reduce environmental pollution or for conservation of natural resources, it may be forced to do so by law. Majority of them felt that the issue of internalising social costs should be carefully considered by companies. Two senior practicing CAs suggested that after carefully determining and valuing social costs, these should be adequately shown in the books of accounts. Companies should also take some steps to mitigate adverse impact of its activities on the society and to compensate society for its wrong actions. They should recognise the fact that in future, they may have to pay for their wrong actions or inaction towards environment. Hence, it is better to accept these costs at an early stage and take voluntary steps to protect the environment.

5.3.8 Environmental Liabilities: A liability is a present obligation of the enterprise arising from past events, the settlement of which is expected to result in an outflow from the enterprise of resources embodying economic benefits in future. [22] Similarly, environmental liabilities are obligations relating to environmental costs. According to ISAR, "Environmental liabilities are obligations relating to environmental costs that are incurred by an enterprise and that meet the criteria for recognition as a liability." [23] In the light of this definition, an environmental liability has been defined in the study as an obligation relating to environmental costs that are incurred by an enterprise and the settlement of which is expected to result in an outflow of resources from the enterprise in future.

Recognition and measurement of environmental liabilities is certainly more problematic because these liabilities deserve special attention. Generally, future obligations are regarded as liabilities, when there is a legal obligation to remedy an existing situation. On account of the inherent unforeseeanability of future legislation, technological changes and the extent or nature of environmental clean-up required, long term environmental liabilities are often difficult to identify and measure. In this sub-section, an attempt has been made to present opinions of the respondents on the recognition and measurement of environmental liabilities.

1. Recognition of Environmental Liabilities: Almost all the CAs (95.1 percent) were of the view that environmental liabilities should be recognised when there is an obligation on the part of the company to incur an environmental cost. In their opinion, these days environmental risk is a significant part of total investment risk. Hence, creditors and investors are interested in knowing significant environmental liabilities of the companies, particularly of manufacturing concerns. Only five CAs felt that environmental liabilities need not be recognised separately in the financial statements. They were of the view that in many cases environmental liabilities are not certain and therefore, at the most, they can be termed as provisions.

About seven-tenths of the responding executives also favoured separate recognition of environmental liabilities in the financial statements. They were of the view that inclusion of these liabilities in the financial statements will give full information to the stakeholders about a company's financial position. This information is very useful for potential stakeholders or creditors to know about credit worthiness of the company and risks involved in their investments.

2. Constructive Obligation: An obligation may be a legal obligation or constructive obligation. Constructive obligation arises when a company is expected to incur a cost because public expects it or it is standard industrial practice or management has declared it in its policy statement. Here enterprise does not have any legal obligation, but only constructive obligation to incur a cost. The respondents were asked whether a company should recognise an environmental liability even when it has only a constructive obligation to incur the cost. They were also requested to specify reasons for their answer.

More than three-fourths of the CAs were in favour of recognising constructive obligations as environmental liabilities. They advanced many reasons for their viewpoint like 'it is a part of social responsibility of business', 'required in broader social interest' and 'required to avoid future penalty'. They were of the view that once a commitment is made by management, it should be fulfilled. Even when management fails to meet it in future, this fact should be disclosed in footnotes to the financial statements together with reasons why this enterprise's management is unable to meet the commitment. Those who were against recognition of constructive obligation as a liability, felt that social obligations need not be recognised as liabilities in the financial statements. Moreover, in the absence of any accounting standard or guideline, it is also difficult to measure these obligations. Their viewpoints on the issue have been summarized in Exhibit 5.12.

Exhibit 5.12: Recognition of Constructive Obligation: CAs' Views

Reasons	Number of CAs
A. For recognizing	
(a) Part of social responsibility	15
(b) Required in broader social interest	7
(c) Beneficial for the company	5
(d) Any other	12
B. For not recognizing	
(a) No accounting standard requires it	5
(b) No need to recognise social obligation in financial statement	7
(c) Any other	6

About three-fourths of the executives were also in favour of recognition of constructive obligation as a liability. They gave many reasons for their viewpoint like it is a 'part of company's social responsibility', shows 'management's commitment towards environmental issues' or 'it is a kind of contractual or moral liability'. However, most of them felt that these obligations should be disclosed in the financial statements only when they can be measured and are material. Otherwise, qualitative disclosures in the footnotes or Director's Report are sufficient. Their viewpoints have been summarized in Exhibit 5.13.

3. Difficulty in Estimating Environmental Liability: Sometimes, it is difficult to estimate the amount of an environmental liability. The respondents were asked whether, in such a case, the liability should not be recognised or a reasonable estimate of it should be provided. Majority of the

CAs and the executives were of the view that in such a case, a reasonable estimate of the liability should be provided in the financial statements. However, the liability should be shown in the footnotes only, instead of including it in the main body of the financial statements. Only a few respondents felt that in these circumstances, the liability should not be recognised in the books of accounts mainly because it is not only difficult but also subjective to provide such an estimate. In rare circumstances, it may not be possible for a company to estimate the environmental liabilities. In such a case, the respondents generally felt that the company can disclose in the financial statements the fact that no estimate has been made, together with reasons for not disclosing the same.

Exhibit 5.13: Recognition of Constructive Obligation: Executives' Opinions

Reasons for Recognizing	Number of Responses
• Part of company's social responsibility	12
• Shows management's commitment towards environment	8
• Moral liability	13
Reasons against Recognizing	
• Difficulty in measurement	2
• Only legal liabilities to be recognized	4
• May not be enforced in future	2

4. Measurement of Long-term Environmental Liabilities: For measurement of long term environmental liabilities, ISAR expresses a preference for using present value of the estimated future expenditures that will be needed, based on the current cost of performing the required activities and existing legal and other requirements. [24] Measuring the liability at full 'current cost' amount is also considered acceptable by ISAR. Majority of the CAs (60.4 percent) were in favour of using 'discounted present value of estimated future expenditure' for measuring environmental liabilities. 24 CAs favoured use of estimated cost of performing the required activities in the current period. They feel that it is difficult to find present value of future expenses and therefore, current cost method is the best. About 45 percent of the executives preferred 'present value method' for measuring these liabilities. However, about one-fifth of them favoured use of 'current cost method' as it is difficult to use the previous method.

5.4 Environmental Reporting

Environmental reporting means incorporation of environmental issues into the annual reports or other statements prepared by corporate entities. It covers the preparation and presentation of information by the management on the environmental activities and performance of a company, for the use of various stakeholders. In the study, environmental reporting has been taken as the disclosure of environment-related information by a company regarding environmental risks, impacts, policies, strategies, targets, costs and liabilities to those who have an interest in such information. This information may be reported through the annual report; a stand-alone corporate environmental report (CER); an environmental statement or some other medium like, video, Internet or staff newsletter.

As mentioned earlier, companies all over the world, are under increasing pressure from the green lobby and regulatory authorities to disclose more environmental information. Recent surveys have shown that a large number of companies (particularly multinational corporations) have already started reporting on the environmental issues. As various stakeholders (including investors, creditors and government) demand greater disclosure of a company's environmental impacts and resulting liabilities, companies will have to take care of this aspect of disclosure more seriously. This Section presents environmental disclosure practices of large manufacturing companies in India as reported by their chief executives. The opinions of the respondents on various issues relating to environmental reporting have also been given here. The Section is further divided into four sub-sections. Sub-section 1 covers general issues relating to environmental information disclosure by companies. Sub-section 2 obtains opinions of the accountants on the appropriate place where this information should be disclosed and type of environmental information to be disclosed. An attempt has also been made to examine practices of the companies regarding environmental disclosures. In sub-section 3, qualitative characteristics that are needed to make environmental information useful for the interested stakeholders have been discussed. Finally, Sub-section 4 lists potential users of these reports.

5.4.1 General Issues: As observed before, though environmental concerns have been heightened in recent years, very few countries have specific financial reporting requirements on environmental disclosures. In the absence of any standard or legal requirements on the issue, many companies do not report on environmental matters. Even those who report on the issue do not follow a particular form and thus, there is no consistency in this reporting. [25] This sub-section presents opinions of the CAs and the executives on some general issues relating to environmental reporting.

1. Uses of Environmental Information: These days, most of the big companies develop internal reporting system to capture the cost effects associated with the adoption of environmental friendly business practices and also the resulting benefits. Companies can generate environmental performance reports for internal decision-making. Moreover, the external environmental reporting can be used to convey the relevant information on environment to a wider audience. In this part, the relevant uses of environmental information have been analysed.

A large majority of the CAs were of the view that environmental information should be used for both the purposes. They felt that this information is required by management in making various business decisions (e.g., capital budgeting decisions and estimation of cost of production). Use of this information by management will make them conscious about environmental concerns. It will not only help a company in complying with environmental laws, but at the same time, will encourage voluntary reporting also. For supporting external reporting of environmental information the respondents again gave various reasons like 'it is a part of good corporate governance practice'; 'this information is required by external stakeholders for decision-making'; 'required for global business' and 'will make accounts transparent'. Only five CAs felt that information should be used for internal decision-making only mainly because external stakeholders are not interested in these figures. On the other hand, eight CAs argued that external reporting of this information is more important and should be given more emphasis.

Regarding actual practices, majority of the companies (75 percent) used environmental information produced by them for both the purposes, i.e., internal decision-making as well as external reporting purposes. 22 executives stated that their companies used most of this information mainly for managerial decision-making purposes and only legally required information was reported to outsiders.

2. Non-disclosure of Negative Environmental Information: It is a common belief that companies operating in environmentally sensitive areas do not disclose environmental information voluntarily, particularly bad news, because the disclosure may jeopardize confidentiality in the sensitive areas and thereby, may adversely affect competitive position of the company. About 64 percent of the CAs agreed with this belief. They felt that fear of action from regulatory authorities and of misuse of information by competitors compel companies not to disclose information on their environmental impacts (particularly negative information). 34

CAs did not consider it to be the most pertinent reason for non-disclosure of environmental information. They were of the view that there are other important reasons also like 'non-sensitivity towards environmental issues', 'lack of top-management's commitment', 'voluntary nature of environmental reporting' and 'absence of accounting guidelines and standards on the issue.'

About three-fifths of the executives felt that the fear of negative publicity is one of the main reasons for non-disclosure of negative environmental information. They were of the opinion that disclosure of bad news (like, oil spills, gas leak, adverse impact on environment) may lead to governmental action and criticism from the public and competitors can take advantage of this. It affects confidentiality in the sensitive areas and may affect competitive position of the company. About 36 percent of the executives were of the view that though this may be one of the reasons for non-disclosure in highly polluting industries; in general, companies do not disclose environmental information because of various other reasons like 'it is not mandatory', 'general public has no interest in environmental figures' or 'lack of accounting standards or guidelines'.

3. Making Environmental Information Disclosure Mandatory: Disclosure of environmental information (except on energy conservation) is not mandatory in India. A review of the existing studies on environmental disclosure practices of Indian companies reveals that in the absence of any accounting standard or legal compulsion, both quality and quantity of environmental reporting has been quite limited. The executives in the study also informed that their companies didn't prepare environmental accounts mainly because it was not required by law. The respondents were asked whether disclosure of environmental information should be made mandatory in India.

About 90 percent of the CAs and 76 percent of the executives were of the opinion that environmental information disclosure should be made mandatory in India. Most of them believed that companies in India are not self-disciplined and very rarely do things voluntarily. Making environmental reporting mandatory will compel these companies to reduce their adverse environmental impacts because they have to report their environmental activities to the public. It will also improve quantity and quality of information reported by the concerns. It will make companies environmentally conscious and will encourage eco-efficiency. If a company has to report its environmental activities or performance to the relevant stakeholders under legal compulsion, the company will definitely try to perform better.

However about 24 percent of the executives were against making environmental reporting mandatory. They felt that as long as a company is taking good care of environment, reporting or non-reporting on it does not matter. Making environmental information disclosure compulsory will unnecessarily lengthen the annual reports, will increase expenditure and lead to wastage of time, cost and energy. Therefore, only voluntary disclosures should be there. Two senior managers suggested, "In the initial years, reporting of some important environmental information items should be made mandatory, that too only for most polluting industries. For other companies, it should only be recommendatory. Later on, in phases, some more items can be selected for mandatory environmental reporting. Moreover, some more companies can be asked to report compulsorily on environment".

5.4.2 Place and Type of Environmental Disclosures: As noted above, majority of the respondents think that a company should disclose information on its environmental activities and performance. This sub-section presents their views on the appropriate place where environmental information should be disclosed and type of environmental information to be disclosed.

1. Place of Disclosure: Environmental information can be disclosed in the annual reports or by some other medium like environmental statements or web-sites. However, annual report is generally considered to be the most appropriate place for reporting this information because, firstly, it has wider audience and secondly, the information given in annual report is verified by auditors. [26] In the annual report also, this information can be disclosed at various places like in the main body of financial statements, in footnotes or in Directors' Report. About 49 percent of the CAs were in favour of disclosure of environmental information in Director's Report or its annexure. Almost same proportion of the CAs supported disclosure of this information in 'separate environmental section' in the annual report. Only 13 CAs felt that this information should be communicated in the main body of financial statements, i.e., Profit & Loss Account or Balance Sheet. These CAs were asked about advantages of disclosing environmental information at this place. They gave two main reasons for their views. Firstly, if the concept of environmental accounting is to be followed seriously, the information on environmental costs, benefits, assets and liabilities should be disclosed in Profit & Loss Account and Balance Sheet. In this way, the matter of environmental accounting will receive due importance. Secondly, disclosure of this information in the financial statements helps in examining the impact of environmental performance on trading results and financial position of the company. This facilitates decision-making by various stakeholders.

Those CAs who were of the opinion that environmental information should not be disclosed in the financial statements also cited various reasons for their viewpoint. Most of them felt that separate disclosure of this information either in the annual report or stand-alone environmental reports will be more beneficial for the interested stakeholders. It will also highlight importance of this information. Other reasons advanced by some of them were, 'difficulty in measurement', 'lack of accounting standards or guidelines' and 'will make annual report bulky'. Exhibit 5.14 summarizes views of the

CAs on the issue.

Exhibit 5.14: Disclosure of Environmental Information in the Main Financial Statements: CAs Views

Reasons	Number of CAs
A. For Disclosing	
Increases seriousness in the issue	3
Helps in checking its impact on trading results and financial position	8
Will be verified by auditors	3
B. For not disclosing	
Difficulty in measurement	7
Lack of accounting standards	5
Will make annual report bulky	17
Not required by law	4
Separate disclosure more beneficial	30
Will lead to confusion	2
Any other	13

Exhibit 5.15 presents practices of the companies regarding place of environmental information disclosure.

The Exhibit shows that about 92 percent of the companies disclosed environmental information in the Director's Report. The executives belonging to 20 companies informed that their companies prepared separate environmental statements also. Most of these companies were public sector corporations or large private sector companies operating in polluting industries. [27] Main body of the financial statements, footnotes or annexure were not found to be popular modes of disclosing details on environmental issues. [28] Two companies made these disclosures at other places also like through Internet and press releases.

As evident from the Exhibit that only one company disclosed information on environmental issues in the main body of the financial statements. The executives were asked why their companies did not disclose information in the main body of financial statements. They gave various reasons for this practice like 'disclosure in the financial statements does not affect usefulness of information for decision-makers', 'will have no effect on conventional accounting format' and 'not recommended by professional bodies'. Some other reasons given by them were, 'will make financial statements bulky', 'may distort purpose of financial statements', 'may create confusion', 'society does not require it' and 'will lead to window dressing'.

Exhibit 5.15: Place of Environmental Information Disclosure: Companies' Practices

Number (Percent)

Place	Companies Disclosing
(a) Main body of the financial statements	1 (1.0)
(b) Footnotes to the financial statements	2 (2.0)
(c) Director's report	92 (91.2)
(d) Separate annexure to the financial statements	3 (3.0)
(e) Separate environmental statements	20 (19.8)
(f) Any other	2 (2.0)

2. Type of Information Disclosed [29]: A company can disclose quantitative information on environment, e.g., information on environmental costs, benefits, assets or liabilities. Similarly, qualitative details on environmental policies and practices of a company can be reported by management. This part seeks to record opinions of the CAs on the type of information which a company should disclose. The actual practices of the companies regarding this have also been examined. Exhibit 5.16 presents comparative analysis of the responses.

The Exhibit depicts that a large majority of the CAs (91.1 percent) recommended disclosure of qualitative as well as quantitative information on environmental issues by a company. Only a negligible number favoured disclosure of only qualitative or only quantitative information mainly to save time and resources of the company.

Exhibit 5.16: Type of Environmental Information Disclosed

Number (percent)

Particulars	CAs' Opinions	Actual Practices
Qualitative only	1 (1.0)	52 (51.5)
Quantitative only	5 (5.0)	2 (2.0)
Both	92 (91.1)	38 (37.6)
N.A.	3 (3.0)	9 (9.0)
Total	101 (100)	101 (100)

Note: N.A. Not Answered

However, more than one-half of the executives informed that their companies disclosed only qualitative details about environmental policies, systems and practices of the company. The annual reports of these companies mainly included information on 'energy conservation' as required to be disclosed under provisions of the Indian Companies Act. *This shows existence of a significant environmental reporting expectation gap in India.* Though users want information on both qualitative and quantitative aspects, most of the companies give only descriptive information on the issue. Only about 38 percent of the responding executives claimed that their companies disclosed quantitative as well as qualitative information.

5.4.3 Qualitative Characteristics of Environmental Information: Qualitative characteristics are very important for the overall credibility of reporting. [30] Reporting on the nature of environmental impacts means that in order to make the processes of measurement and reporting feasible and credible, certain underlying assumptions may need to be made about either the reporting enterprise or the data being reported on.[31] Most of the underlying assumptions and qualitative characteristics of financial reporting are equally applicable to environmental reporting also. For making environmental report useful for external stakeholders, it must also possess certain characteristics like it should be relevant, reliable and understandable. There is a need to present data in as uniform manner as possible, if it is to be of any significant use to external shareholders. However, application of some characteristics is a bit complex in environmental reporting. For example, the application of materiality concept in case of environmental reporting is heavily dependent on the carrying capacity of the receiving environment (such as availability of landfill capacity or background air pollution levels). Moreover, environmental reports may contain some information, which cannot be objectively determined or physically quantified, e.g., environmental impact of company's activities and long-term sustainability. Accordingly, some qualitative characteristics may have to be modified. The standard setters should make adequate provisions accordingly while setting conceptual framework for environmental reporting.

The respondents were asked about their opinions on the qualitative characteristics that must be possessed by the reported environmental information to make it more beneficial for the users. 'Useful in decision-making' was considered as the most important characteristics of environmental information by about 35 percent of the CAs. According to them, there is no use of giving any information in the annual reports or by some other medium, unless it is useful to the stakeholders in making various financial or other decisions. Therefore, only material and significant information that facilitate decision-making should be supplied by a company. Objectivity was another characteristic considered important by them. Till date, in the absence of any accounting standard or guideline on the issue, the environmental information given in the annual report by companies has been generally subjective (e.g., only positive or good information is generally given). In order to make environmental information more useful for the interested external stakeholders, it should be objective. Other characteristics considered as the most important by some of them were relevance, reliability, understandability and completeness. Some CAs felt that all these characteristics are equally important as there is an overlapping in some of these characteristics. For example, in order to be useful in decision-making, information is to be relevant, reliable, objective and complete.

Majority of the executives (55.4 percent) were also of the view that environmental information supplied by a company should be useful in decision-making. Most of them felt that in order to make this information valuable for the stakeholders, it is essential that only relevant and objective information is provided to them. Other important characteristics most preferred by about 45 percent of the executives were understandability, consistency, completeness and reliability. Again some managers were of the view that all these characteristics are important and must be taken care of while preparing environmental reports.

It is to be noted that least important characteristics according to the qualified respondents were timeliness and comparability. They were of the opinion that sometimes it may not be possible for a company to give timely details on environmental issues, as it is difficult to measure its impact in a particular time frame. Moreover, in some circumstances, it may become relevant later on due to changes in legislation or otherwise. Similarly, environmental information given by various companies may vary due to differences in their product, process, technology, design or legal requirements for various industries. Thus, certain features of environmental disclosures made by them cannot be compared.

Cost-effectiveness of Environmental Information: A company before undertaking any activity generally conducts cost-benefit analysis. Most often, this is one of the bases for making many important decisions. The cost-benefit principle says that cost of applying something should not exceed its benefit. In financial reporting also one of the pervasive constraints is that the benefit to be gained from providing information should be greater than the cost of providing it.

[32] Majority of the CAs (80.2 percent) agreed that the benefits of providing environmental information should be more than the cost involved in it. In their opinion, a formal cost-benefit analysis will help a company in justifying its decision relating to provision of environmental information to stakeholders. Some of them clarified, "It does not mean that to save cost, no or very little information should be given to the users. A certain minimum level of information absolutely essential in decision-making should be provided". 20 CAs were of the opinion that in providing environmental information, this kind of cost-benefit analysis should not be done. They gave two main reasons for their viewpoint. Firstly, disclosure of environmental information is a part of social responsibility of a company and in these matters, cost factor should not be considered. Secondly, in case of environmental issues, though cost of doing something is generally tangible, most of the times, the benefits are intangible (e.g., enhanced image due to environmental reporting). Moreover, cost of not taking an action (e.g., loss of goodwill due to non-disclosure of environmental information) is also generally difficult to measure. It is, therefore, not feasible to conduct this kind of analysis.

About seven-tenths of the executives informed that the benefits of environmental information to the potential users are generally greater than the cost of providing such information. They told that these days most of the companies prepare environmental statements for internal decision-making purposes. Since this information is already available in the organization, its reporting to the external stakeholders does not involve much cost. On the other hand, various stakeholders (including creditors, government and investors) can take benefit of this information in making relevant financial or other decisions. The company is also benefited because of enhanced image, increased sales and reduced pressures from government.

5.4.4 Users of Environmental Information: The IASC

conceptual framework differentiates between seven user groups, but is based upon a primary assumption that the needs of financial investors are pre-eminent, and that if their needs are met, then the needs of most other users should be at least partially met. There are various stakeholders like investors, creditors, government and society at large who are interested in environmental disclosure made by a company. United Nations working Group of Experts, ISAR identified various users of environmental information like shareholders/ prospective owners, internal line managers, local/regulatory authorities or environmental organizations. [33] These users generally have varying information needs. A company has to identify its relevant users and accordingly decide about the type and nature of environmental information to be disclosed. This also affects the mode by which this information is to be reported and the place where it is to be disclosed (e.g., generally, information to government is given in statutory forms and to shareholders in the annual reports). In this part, the CAs and the executives were asked to rank the given users of environmental information in order of importance.

'Society at large' was considered as the most relevant user of environmental information by about 58 percent of the CAs. 'Government' and 'Environmental Organisations/NGOs' were two other users considered important by them. These CAs were of the opinion that environmental information is mainly useful for the society at large. All the constituents of society need the details on environmental performance of an organisation because all of us are a part of environment and in one way or the other, are affected by it. Thus, this information should be directed to a wider audience. In addition, specific purpose statements can be prepared to meet specific needs of some segments of the society. It is to be noted that a significant percentage of the CAs (more than three-fifth) did not consider suppliers, competitors, customers and journalists to be the relevant users of this information (they gave zero ranks to these users).

About three-fourths of the executives considered 'government' as the most important user of environmental information. One of the manager informed, "Generally environmental information is published to demonstrate to the regulatory authorities that the company is complying with environmental laws and also taking voluntary steps to protect the environment. Most of the times, company is legally required to submit these details to the government. Environmental reporting helps in releasing pressures from regulatory authorities. Hence, government/regulatory authorities are the most important users of environmental information". Society at large and environmental organizations (NGOs) were two more users considered important (given one of the first three ranks) by more than 65 percent of the respondents. According to them, environmental reporting helps to build the trust of the local community. It makes it easier for a company to operate and expand. Furthermore, in recent years, non-governmental organizations have become quite active towards environmental problems and many a times their efforts lead to judicial action. The least important users according to the executives were - competitors, suppliers and customers. Some companies informed about other users also like foreign institutional investors.

5.5 Environmental Audit

The term environmental audit has been explained by various authors and professional bodies differently. [34] Broadly, it involves audit of environmental management system, environmental activities, and environmental performance of an organisation. But for this study, environmental audit has only been taken as audit of information given in the annual reports or environmental statements prepared by a company. Thus, the scope of environmental audit is restricted to only verification of environmental information reported by a concern. It does not, therefore, covers various technical and other aspects involved in environmental audit like energy audit, procedure audit, emission audit or environmental performance audit. Environmental audit has been defined as a systematic process of objectively obtaining and evaluating the evidence relating to performance of an organisation as reflected in the environmental statements. Thus, environmental audit attempts to examine and report verifiable, quantitative and qualitative information in the relevant areas.

Some existing research works on environmental reporting have highlighted the growing need for verification of environmental reports produced by a concern. [35] This part seeks to analyze views of the responding accountants on the need and significance of environmental audit. In addition an attempt has been made to examine practices of the companies relating to preparation of environmental statements and their verification.

5.5.1 Need for Audit of Environmental Information: The views of the CAs on the need for audit of environmental statements were obtained in the study. A large majority of the CAs (91.2 percent) were of the opinion that environmental reports should be audited. They felt that in order to make environmental information given by a company objective, reliable and useful in decision-making, it should be verified by an independent auditor.

The CAs who advocated verification of environmental report were further asked to rank given benefits of environmental audit in order of importance. Their responses are presented in Exhibit 5.17. It shows that about 41 percent of the CAs considered 'increases credibility of information provided' as the most important benefit of environmental audit. About three-fourths of them gave one of the first three ranks to this benefit. 'Ensures compliance with environmental laws' was another benefit given one of the first 3 ranks by about 84 percent of the CAs.

Only nine CAs did not find any need for this verification. One senior Chartered accountant argued, "At present, only a few companies provide information on environment. Most of them provide only qualitative information on environment in the Director's Report.

Exhibit 5.17: Benefits of Auditing Environmental Statements: CAs' Opinions

Number (Percent)

Benefits	Ranks			
	0	**1**	**2**	**3**
(a) Increases creditability of information provided	4 (4.4)	41 (45.6)	16 17.8)	11 12.2)
(b) Guarantees validity and accuracy of environmental data and reports	8 (8.9)	19 (21.1)	25 (27.8)	12 (13.3)
(c) Ensures compliance with environmental laws	4 (4.4)	33 (36.7)	21 (23.3)	22 (24.4)
(d) Checks adherence with pre-determined policies of management	16 (17.8)	10 (11.1)	8 (8.9)	15 (16.7)
(e) Assists management in decision-making	17 (18.9)	10 (11.1)	4 (4.4)	9 (10)
(f) Helps in responding to pressures from regulatory authorities and NGOs	11 (12.2)	10 (11.1)	5 (5.6)	7 (7.8)
(g) Any other	-	-	-	-

Contd...

Benefits	Ranks		
	4	**5**	**6**
(a) Increases creditability of information	10 (11.1)	6 (6.7)	2 (2.2)

provided			
(b) Guarantees validity and accuracy of environmental data and reports	15 (16.7)	8 (8.9)	3 (3.3)
(c) Ensures compliance with environmental laws	4 (4.4)	4 (4.4)	2 (2.2)
(d) Checks adherence with pre-determined policies of management	15 (16.7)	18 (20)	8 (8.9)
(e) Assists management in decision-making	23 (25.6)	20 (22.2)	7 (7.8)
(f) Helps in responding to pressures from regulatory authorities and NGOs	5 (5.6)	15 (16.7)	37 (41.1)
(g) Any other	-	-	-

Hence, there is no need to audit it". Three of them gave non-mandatory nature of environmental audit as the main reason for not recommending this audit. They suggested that first of all, all the companies should be encouraged to report on environment voluntarily. Then after some years, reporting of externally audited environmental information can be made mandatory. Other reasons given by some of them were 'time consuming', 'expensive', and 'no verification standards'.

5.5.2 Preparation and Audit of Environmental Statements by Sample Companies: Exhibit 5.18 presents practices of the sample companies regarding preparation and audit of environmental statements.

The Exhibit depicts that about seven-tenths of the responding companies prepared environmental statements. These companies were further asked whether preparation of environmental statements was statutorily required in their industries. The Exhibit shows that in large majority of these companies (92.8 percent), preparation of environmental statements was a statutory requirement. [36] 56.5 percent of these companies also prepared environmental statements voluntarily. Generally a high proportion of public sector companies prepared these statements. Their commitment towards betterment of society and open disclosure policies might be the reasons for that.

Exhibit 5.18: Preparation and Audit of Environmental Statements: Actual Practices

Number (percent)

Particulars	Yes	No	N.A	Total (100 percent)
1. Preparation of environmental statements	70 (69.3)	27 (26.7)	4 (4)	101
2. Preparation of mandatory environmental statements	64 (92.8)	5 (7.2)	-	69*
3. Preparation of voluntary environmental statements	39 (56.5)	30 (43.5)	-	69*
4. Audit of environmental statements	48 (69.6)	21 (30.4)	-	69*

*Note: These questions were applicable to 70 executives but answered by 69 executives only.

It was further noted that about seven-tenths of the companies produced audited environmental statements. The executives from these companies were requested to rank the given benefits of auditing these statements in order of importance. Exhibit 5.19 reveals that 'ensures compliance with environmental laws' was considered as the most important benefit of this kind of environmental audit by about 73 percent of the executives. According to them, verification ensures that the company is complying with these laws and if any default is there, it comes to the knowledge of the management. About three-fifths of the managers felt that environmental audit 'increases credibility of information provided' to different stakeholders. Some executives gave other benefits also like 'it helps in self-appraisal', 'checking of information before submission to authorities' and 'ensures that correct information is supplied to authorities'.

Exhibit 5.19: Benefits of Environmental Audit: Executives' Perception

Number (Percent)

Benefits	Ranks					
	0	1	2	3	4	5
(a) Increases credibility of information provided	10 (20.8)	30 (62.5)	2 (4.2)	3 (6.3)	2 (4.2)	1 (2.1)
(b) Guarantees validity and accuracy of data	16 (33.3)	19 (39.6)	8 (16.7)	2 (4.2)	3 (6.3)	-
(c) Ensures compliance with environmental laws	4 (8.3)	35 (72.9)	2 (4.2)	3 (6.3)	2 (4.2)	2 (4.2)
(d) Assists management in decision-making	27 (56.3)	8 (16.7)	2 (4.2)	4 (8.3)	4 (8.3)	3 (6.3)
(e) Helps in responding to pressures from governmental or non-governmental authorities	31 (64.6)	7 (14.6)	2 (4.2)	1 (2.1)	2 (4.2)	5 (10.4)
(f) Any other	44 (91.7)	2 (4.2)	1 (2.1)	1 (2.1)	-	-

About 30 percent of the companies produced unaudited environmental statements. These companies were enquired about the problems in verification of these reports. Exhibit 5.20 gives analysis of the responses. [37] A large majority of the executives (93.8 percent of the companies which did not produce verified environmental reports) stated that their companies did not produce verified statements because it was not required by law. Some of them were of the opinion that environmental audit should be made mandatory in India (particularly in environmentally sensitive industries). 'No specific auditing standards' was another problem considered important by the companies. These executives felt that in the absence of specific standards on the issue, the verification of these statements is meaningless. It will not add much incremental value to these statements

Exhibit 5.20: Problems in Environmental Audit: Executives' Views

Number (Percent)

Problems	Ranks					
	0	1	2	3	4	5
(a) Expensive	24 (75)	-	1 (3.1)	-	7 (21.9)	-
(b) Time consuming	24 (75)	-	-	7 (21.9)	-	1 (3.1)
(c) Not mandatory	1 (3.1)	30 (93.8)	-	1 (3.1)	-	-
(d) No specific standards	19 (59.4)	3 (9.4)	9 (28.1)	-	1 (3.1)	-
(e) Any other	30 (93.8)	2 (6.3)	-	-	-	-

5.5.3 Environmental Auditors: The CAs who were in favour of providing verified environmental information to the interested stakeholders were further asked to render their opinions on the number, status and qualifications of the environmental auditors. 93.7 percent of the CAs [38] preferred appointment of a team for this task. They were of the opinion that environmental audit involves a lot of things and an individual may not handle all these aspects. Only six CAs were in favour of giving this task to an individual. Some of them argued, "An individual can easily handle this task because till date, Indian companies provide only limited information on environmental issues. Hence, appointment of a team for this purpose will lead to wastage of energy, time and cost". Some of these CAs were of the opinion that an external auditor should be given this task as it will increase credibility, validity and objectivity of environmental information provided by a concern.

The CAs who favoured appointment of a 'team' of environmental auditors were further asked about the preferred qualifications of the persons in this team. Majority of the CAs (75.8 percent) were in favour of appointing 'qualified environmental auditors' for auditing environmental reports. They were of the view that environmental audit, being a new area, requires separate education and training and only qualified environmental auditors can perform this task efficiently. More than one-half of them supported appointment of 'chartered accountants' for verifying the financial details given in the report. About 37 percent wanted 'engineers' in the team for handling technical matters.

Analysis of Actual Practices: An analysis of actual practices shows that a large majority of concerns (95.2 percent) appointed a team to do environmental audit. [39] Environment is a multidisciplinary subject and an individual cannot handle the task of environmental audit effectively. Hence, a team was given this task.

The executives were further asked about the status of the environmental auditor (internal vs. external) appointed by them. Most of these companies (25) appointed external persons as environmental auditors. [40] These companies found it necessary to ensure independence and objectivity in this task. One of the executives informed, "Environmental risk is an important part of credit risk and therefore, creditors want independently verified environmental statements from the companies to get authentic information about their environmental performance, potential hazards and future environmental liabilities."

However, some companies (13) gave this task to internal auditors. They gave two main reasons for this practice. Firstly, environmental audit is not a statutory audit and therefore, it is not necessary to appoint external persons for this. Secondly, internal auditors are available within the organization. They know about company's policies and procedures. Hence, it is better to appoint them as environmental auditor. The executives were also asked to tell about composition of environmental audit team. 81.7 percent of the companies appointed 'qualified environmental auditor' to do this work. In some companies 'engineers' and 'chartered accountants' were employed to handle specific technical and accounting aspects respectively. Some executives felt that the 'chartered accountants' are able to provide input with regard to the verification of financial data, cost-benefit analysis, compliance status and implementation of EMS.

5.6 Summary and Conclusions

This Section presents summary and conclusion of the primary data analysis. The main objective of this Chapter was to examine existence of an environmental accounting and reporting (EAR) expectation gap in India. For this purpose, opinions of the selected Chartered accountants were obtained regarding their expectations about corporate environmental accounting and reporting in India. In addition, the actual EAR practices of the selected large manufacturing companies operating in India were analysed with the help of primary data to examine whether they fulfil expectations of the users of environmental information.

The qualified accountants and the senior executives in large manufacturing companies felt that environment is a big challenge for business these days. The CAs, in general, were of the opinion that companies should establish EMS to meet these challenges. Establishment of a good EMS helps in maintaining clean and green environment around the factories, controlling pollution emission and thereby meeting legal standards in this area. An examination of actual practices showed that a large majority of the companies (89.11 percent) already had EMS in their organizations, while some companies were planning to introduce it in near future.

A large majority of the respondents were of the opinion that the current financial accounting framework is not comprehensive enough to deal with environmental problems. Most of them were in favour of development of a separate conceptual framework and a specific comprehensive standard for environmental accounting to effectively deal with specific issues involved in it.

Almost all the CAs felt that a company should use environmental accounting system mainly because it helps in complying with environmental laws. The other two benefits regarded important by them were, 'helps management in decision-making' and 'provides better estimates of cost of producing a product'. But it was found that a substantially high proportion of the companies (93.1 percent) did not prepare environment related accounts for external reporting purposes. Thus, a significant expectation gap exists in this area. The managers gave non-mandatory nature of environmental accounting as a justification for not preparing these accounts. It seems that the companies in India are reluctant to take initiatives in this regard. Their main aim seems to be furnishing only that information which is legally required or which can influence the decisions of their stakeholders about them. The executives also mentioned 'no accounting standard and guideline' as the reason for not preparing these accounts. It was interesting to note that more than one-half of the companies had practice of generating and using internal environmental statements for managerial decision-making. It seems that companies have started environmental accounting in providing inputs for successful implementation of EMS. The study found popularity of the concept of environmental accounting in India only at the management accounting level.

The respondents were of the opinion that environmental cost information generated by environmental accounting is useful in 'designing various processes using environmental friendly technologies'. They also found this information useful in appraisal of investment for environmental risks and environmental impact assessment. For treatment of environmental costs and benefits in the financial statements, both CAs and the executives recommended use of same financial accounting principles with some modifications according to the nature and extent of environmental costs.

The CAs felt that a company should try to determine external adverse impacts of its activities (social costs). They recommended consideration of these costs in strategic decision-making. About 88 percent of them were of the view that a company should also try to value these social costs so that it can take necessary steps to reduce adverse impact of its activities on environment. Regarding actual practices, though majority of the companies tried to determine external adverse impact of their activities on environment (social costs), most of them did not try to value these social costs. About 90 percent of the companies considered these costs in making strategic business decisions like, in capital budgeting. A large majority of the CAs and the responding executives were of the view that due to growing legal and social pressures, most of these external costs will become internal in future (e.g., companies will have to bear penalties or clean up pollution).

The respondents generally supported recognition of environmental liabilities in the financial statements when there is an obligation (legal as well as constructive) on the part of a company to incur an environmental cost. When there is difficulty in estimating an environmental liability, a reasonable estimate of it should be provided. In rare circumstances, when it is not possible for a company to estimate environmental liabilities, this fact along with the reasons therefore, should be disclosed. For measurement of long-term environmental liabilities, most of the respondents preferred use of 'discounted present value of expected future costs using risk-free rate'.

The CAs, in general, recommended that a company should use environmental information for both internal decision-making and external reporting. They also felt that companies must disclose qualitative as well as quantitative information on these issues. Most of them preferred disclosure of this information in separate environmental statements to 'highlight its importance' and to 'avoid any confusion'. Furthermore, they also approved disclosure of qualitative information in the 'Director's Report'.

An examination of the actual practices revealed that about one-half of the companies disclosed only qualitative information on environment in the annual reports. The companies mainly used the Director's Report to supply environmental information to external stakeholders. [41] The study found existence of a significant environmental reporting expectation gap in India. Some companies (about one-fifth) also prepared separate environmental statements to communicate this information. Some of the executives were of the view that disclosing these details in the main body of financial statements would serve a better purpose. However, more than one-half were against this as it may 'make financial statements bulky', 'create confusion' and 'will not facilitate decision-making by stakeholders'.

'Usefulness in decision-making' was considered as the most important characteristic of good environmental statements by majority of the CAs and the executives. For this, the information should be relevant and objective. Timeliness and comparability were considered less important by the executives. Majority of the respondents were of the view that benefits of environmental information to the potential users are greater than the cost of providing it.

The environmental statements should be prepared keeping in mind users' needs. 'Society at large' was considered to be the most relevant user of environmental information by the CAs. 'Government', 'environmental organisations' and 'owners' were other relevant users, according to them. However, the executives found 'government' to be the most important user of this information because these statements were mainly prepared to comply with the legal requirements. It helped in reducing pressures from regulatory authorities. The least important users according to the respondents were - competitors, suppliers and customers.

A large majority of the CAs were in favour of mandatory disclosure of environmental information. About three-fourths of the executives were also of the same view. It seems that compulsory environmental information disclosures will increase seriousness in the issue and will lead to overall increase in environmental reporting by Indian companies. [42]

The findings of the study revealed that a large majority of the CAs favoured verification of environmental statements prepared by a company mainly to 'increase credibility of information provided' in them. They felt that this audit would also help a company in complying with environmental laws. The responding CAs felt that a company should submit its environmental report to independent scrutiny mainly to underpin the credibility of the environmental report. [43] The CAs, in general, favoured appointment of a team, mainly comprising of external environmental auditors to do this task. They felt that 'Chartered accountants', 'engineers' and 'cost accountants' might also form a part of this team to handle specific tasks.

A scrutiny of actual practices followed by the large Indian companies revealed that while most of the companies prepared statutorily required environmental statements, only about 39 percent prepared these statements voluntarily. About seven-tenths of these companies got these statements verified by an auditor mainly to ensure compliance with environmental laws and to increase credibility of the information provided in these reports. The task of environmental information audit was generally given by the responding companies to a team of auditors. Most of these companies appointed external environmental auditors for this purpose; while some companies assigned specific tasks to engineers and chartered accountants. However, about three-tenths of the companies produced unaudited environmental statements mainly because verification of environmental reports was not required by law. It is therefore, recommended that Environmental information audit (EIA) should be made mandatory at least in major polluting industries. This can be extended to other industries, after the audit systems in the polluting industries are well documented. 'No specific standard' on the issue was another problem considered significant by the companies. It was felt by the respondents that to increase credibility and reliability of environmental information supplied to various interested stakeholders, it should be verified by some independent auditors as per accounting standards issued by the ICAI.

The findings of the study reveal that the respondents were convinced by the need for having an EAR system in the large companies in India. No significant difference was found between the views of users and the preparers on various issues raised in the study. Both the CAs and the executives found usefulness of environmental information in strategic decision-making. They felt that benefits of providing environmental information to the relevant users are more than cost of providing this information. The CAs and the executives favoured mandatory environmental reporting in India. But when actual EAR practices (as reported by the executives) were analyzed, significant differences were found in expectations of the users and actual practices of the companies. Though, a large majority of CAs felt that companies must adopt environmental accounting system, only a negligible number of companies were found to be preparing these accounts for external reporting purposes. The users preferred separate environmental statements and Director's reports for disclosing qualitative as well as quantitative information on environment. However, most of the companies had the practice of disclosing only qualitative environmental information in the annual reports mainly through the Director's Report. Indian companies failed to provide adequate disclosures of their environment activities and performance. The evidence is consistent with the existence of an environmental accounting and reporting (EAR) expectation gap within India.

Notes

1. Refer to Chapter IV, for review of existing literature on EAR.

2. The power generation and mining industries were not excluded as they were considered to be 'the most polluting' industries under Environment (Protection) Rules, 1992. These Rules give a list of industries which require consent under any Act relating to environment. The accounting and reporting on environment is important for these companies because they operate in environmentally sensitive areas. See Appendix II for the list of 17 Most Polluting Industries in India.
3. Dey N.B., 'Environmental Auditing for Effective Corporate Management', in Pramanik A.K., *Environmental Accounting and Reporting*, Deep and Deep Publications Pvt. Ltd., New Delhi, (2003), p. 55.
4. EPA, *'An Introduction to Environmental Accounting as a Business Management Tool: Key Concepts and Terms'*, U.S. EPA Primer, Washington, DC: EPA, (1995a), p. 26.
5. It is generally believed that environmental accounting helps a company in planning cost reductions. According to EPA, "An important function of environmental accounting is to bring environmental costs to the attention of corporate stakeholders who may be able and motivated to identify ways of reducing or avoiding these costs while at the same time improving environmental quality." *Ibid*, p.18.
6. To quote EPA "The goal of environmental accounting is to increase the amount of relevant information that is made available to those who need or can use it. The success of environmental accounting does not depend on correctly classifying all the costs that a firm incurs", EPA, *op. cit.*, p. 12.
7. The survey of ISAR revealed that in the absence of any formal instructions from regulatory authorities, enterprises used their discretion in maintaining environmental cost records. See Rathore S., *International Accounting*, Prentice Hall of India Pvt. Ltd., New Delhi, (1996), p. 421.
8. UN-ICAI Seminar Report, *'Environmental Accounting and Reporting'*, conducted at New Delhi, (Aug. 2-4, 1999), p. 5.7.
9. Their thinking is contrary to the definition of assets given by the ICAI. The ICAI defines assets as, "tangible objects or intangible rights owned by an enterprise and carrying probable future benefits." See The Institute of Chartered Accountants of India (ICAI), Guidance Note on *Terms Used in Financial Statements*, New Delhi: ICAI, (Sept. 1983), p. 8.

10. UN-ICAI Seminar Report, *op. cit.*, pp.5.7-5.9.
11. FASB (USA) defines assets as, "probable future and economic benefits obtained or controlled by a particular entity as a result of past transactions or events." See FASB, *Elements of Financial Statements of Business Enterprises*, Concept No. 3, Stamford, FASB (Dec. 1980), Para 19.
12. One CA suggested, "In the concepts of assets as applicable to environmental accounting, the term carrying 'probable future benefits to the enterprise' should be replaced by carrying 'probable future benefits to the society'."
13. Refer to Appendix III for Carbon Trading.
14. The term' prior period items', refers only to income or expenses which arise in the current period as a result of errors or omissions in the preparation of the financial statement of one or more prior periods. The term does not include other adjustments necessitated by circumstances which though related to prior periods, are determined in the current period. Refer to Accounting Standard 5 issued by the ICAI.
15. UN-ICAI Seminar Report, *op. cit.*, p. 5.11.
16. INFORM (a not for profit group in USA) in two of its reports based on studies conducted in1985 and 1992 showed that significant environmental benefits (in the form of cost savings) can be derived by using pollution prevention projects (INFORM, INFORM, "Cutting Chemical Wastes", (1985); INFORM, "Cutting more Chemical Wastes," (1992), New York, adopted from EPA, (1995a), *op. cit.*
17. Hanson D.R. and Mowen M.M., *Cost Management Accounting and Control,* Chapter 17 – Environmental Cost Management, South Western College Publishers, Cincinnati, USA, 3rd edition, (2000), pp.665-698.
18. It is to be noted here that the study talks about only direct measurable environmental benefits to the company from incurring environmental costs. Indirect/social/ intangible benefits have not been covered in the study.
19. One senior practicing CA from Delhi proposed, "A company should try to estimate (even if approximate) the likely benefits over a period of time and should calculate their present value by discounting them at an appropriate cost of capital. This present value represents the value of the future benefits of environmental costs incurred by a company."

20. One of the limitations of the present study is that it does not deal with recognition and measurement of social costs and benefits.
21. For example, an executive in a cement company informed, "The company checks the impact of fumes and cement dust generated on the health of its employees. Then, it tries to determine the average amount spend by an employee to maintain his health. Accordingly, the company allocates money for good health, food, medical allowance and treatment of employees." Another senior executive working in a highly_polluting chemical plant told, "Various decisions like tree plantation, voluntary pollution prevention, and health and safety of employees, are taken by the company on the basis of a rough estimate of value of the adverse impact of company's activities on environment and health of the employees."
22. According to FASB, "Liabilities are probable future sacrifices of economic benefits arising from present obligations of a particular entity to transfer assets or provide services to other entities in the future as a result of past transactions or services." See, Financial Accounting Standard Board, Concept No. 3, *Elements of Financial Statements of Business Enterprises*, Stamford: FASB (Dec. 1980), p. 12.
23. UN-ICAI Seminar Report, *op. cit.*, p. 5.12.
24. UN-ICAI Seminar Report, *op. cit.*, p. 5.18.
25. Refer to Section 4.2 for existing studies on environmental reporting.
26. Tilt in a study conducted in 1994 found that annual report was considered by the companies to be the most appropriate medium for disclosing information on environment, demonstrated by their preference for it over other mediums. See Tilt A., "The Content and Disclosure of Australian Corporate Environmental Policies", Source: www.ssh.flinders.edu.au/commerce/researchpapers/aa.4.htm, last accessed on 30.1.2005.
27. In a survey of Top 100 companies in 1997, Clarks and Gibson also found that companies in sectors with high environmental impacts had a tendency to produce higher proportion of environmental reports (See Clarks J. and Gibson M., "Use of Corporate Social Disclosures in the Management of Reputation and Legitimacy: A Cross Sectoral Analysis of UK Top 100 Companies", *Business Ethics– A European Review*, Vol. 18, No.

1, (Jan. 1999), pp. 5-13).

28. Jones (2000), in a survey of Annual/Financial reports of 88 EU companies also discovered that two-thirds of the companies disclosed nil or limited information in the financial reports. See Jones K., '*Study on Environmental Reporting by Companies*', Commission of European Communities, DG Environment, Centre for Environmental Informatics, University of Sunderland, (March 2000).
29. The questionnaire for CAs consisted of one question on the environmental information to be disclosed by a company. An 'Index of environmental disclosure' was constructed listing the items of environmental information that a company can disclose in the annual report or environmental statement. The Index consisted of 23 items. The CAs were requested to evaluate each of these items independently and assign them points on the basis of their importance and usefulness in making sound investment and other decisions and in evaluating environmental performance of a company. This Index was used to evaluate environmental disclosure practices of the sample companies in their annual reports. Therefore, the analysis of this question has been included in the Chapter VI.
30. FASB's Statement of Financial Accounting Concept (SFAC) No. 2 examines the characteristics that make accounting information useful. For details on hierarchy of Accounting Qualities, refer to Porwal L.S., *Accounting Theory: An Introduction*, Tata McGraw-Hill Publishing Company Limited, New Delhi, 3rd ed. (2001), p. 117.
31. Federation Des Experts Comptables Europeans (FEE), '*Discussion Paper towards a Generally Accepted Framework for Environmental Reporting*', (1999), Source: http://www.fee:brussess. fee.be/publications/main.htm.
32. Porwal, 2001, *op. cit.*, p. 139.
33. UN-ICAI seminar report, *op.cit.*
34. Refer to Section 2.4.1 for these definitions.
35. See for example, KPMG, *International Survey of Environmental Reporting*, (1999) and KPMG, *International Survey of Corporate Sustainability Reporting*, (2002).
36. Rule 14 of the Environment (Protection) Act, 1986 (as introduced on 13.3.92) requires certain industrialists to submit environmental audit report every year in prescribed format to

SPCB. It is mandatory in case of any industry requiring consent under any Environment Act.

37. This question was applicable to only 21 companies which produced unaudited environmental statements. But 32 executives talked about problems in getting these statements verified. It seems that 11 executives (those who didn't answer the previous question) talked about problems of environmental audit in general. Their views have also been included in the analysis.
38. The question was answered by 95 CAs.
39. In the previous sub-section, only 48 companies informed that they produced audited environmental statements. But here 62 companies talked about appointment of environmental auditor. It appeared that some of them were informing about appointment of environmental audit in general or for obtaining ISO 14001, where it is an essential part.
40. Only 38 companies provided this information.
41. *Roy (2000)* also observed that the common practice followed by the Indian companies in both public and private sector over the years regarding environmental disclosure was to offer descriptive information and itemize the same in the Director's Report. Only a few companies disclosed information through notes, schedules and a separate social account. *Jaffer et al. (2002)* also found most of the environmental information in the Review of Operations section and in Chairman's statements. As compared to this, *Jones (2000)* in a study of 88 European companies for environmental disclosures found that out of 13 sections identified in the annual reports, three-fourths of the companies provided this information in the 'environmental section'.
42. *Frost (2001)* found a significant improvement in overall reporting by Australian companies on their environmental performance after introduction of mandatory reporting in 1998. However, *Larrinaga (1999)* found otherwise in a study on evaluation of mandatory environmental accounts in Spain. He did not found any improvement in environmental reporting after introduction of mandatory environmental reporting requirement.
43. As compared to this, the research published in 1996 by the Global Environmental Management Initiative (GEMI) and the Investor Responsibility Research Centre (IRRC) in Washington

DC found that the significant stakeholder groups thought that verification added little, if anything, to the credibility of the reports overall. The lack of standards for verification was found to be the main drawback at that moment (*GEMI 1996*).

6

Environmental Disclosure Practices of Select Companies

It was observed in the previous Chapter that the executives were well aware of the growing need of incorporating environmental issues in the financial statements. Though the companies generally did not prepare environmental accounts for external reporting, most of them used this information in internal decision-making. The managers belonging to almost all the companies told that their companies report on the environment. But more than one-half of the companies gave only qualitative information mainly in the Director's Report or its annexure. The disclosure of quantitative/financial information was not common. The executives recognized the need for a proactive approach to environmental management. It was felt that to meet information needs of various stakeholders (including Government, NGOs and society at large), companies should publish relevant, objective and useful information on the issue.

This Chapter presents an analysis of published annual reports of the selected companies to examine their disclosure practices regarding environmental issues. The main hypothesis to be tested here is that companies in India do not give elaborate reporting on environmental issues. In addition, an attempt has been made to statistically test the combined influence of the selected parameters on the environmental reporting practices of the sample companies with the help of multivariate analysis. The Chapter is divided into six Sections. In Section 6.1, scope of the study, hypotheses, and methodology used have been described. Section 6.2 evaluates extent of environmental information disclosed by the companies in the annual reports. It also examines type, length and location of the disclosures in these reports. Section 6.3 presents analysis of environmental disclosure scores (EDS) obtained by the companies. Section 6.4 introduces regression analysis and the measures to test the goodness of fit for use in this Chapter. This Section also explains building up of model for the present study. Section 6.5 analyses the results of the estimated regression equations. Finally, Section 6.6 summarizes the Chapter.

6.1 Scope and Methodology

This Section explains the scope and the detailed methodology used in analysing environmental disclosure practices of the selected companies. Sub-section 1 explains the construction of an Index of environmental disclosure. While sub-section 2 deals with the methodology used in the analysis of annual reports, sub-section 3 talks about the multivariate analysis used in the present Chapter.

6.1.1 Construction of an Index of Environmental Disclosure: A company can disclose information on various environmental issues in its annual report or environmental statement. In the present study, an attempt was made to construct an 'Index of environmental disclosure', listing the items of environmental information that a company can disclose in the annual report or environmental statement. The Index consisted of 23 items.

Many sources [1] such as, books on accounting, social responsibility reporting, environmental accounting and reporting; views and recommendations of various professional accounting bodies and committees on social and environmental reporting; and articles published in this area were consulted in determining the items for inclusion in the 'Index'. A small survey of the annual reports was also made for this purpose. These items were classified under seven major groups. The rationale of such grouping lies in the fact that a decision-maker/user of environmental information cannot evaluate the environmental performance of a company directly from individual items. For example, environmental accounting practices of a company cannot be judged merely from the information relating to environmental costs in the annual reports. For this, information on environmental assets, liabilities, and significant accounting policies is required.

Similarly, an organisation's efforts on conservation of natural resources can be evaluated by information on statutorily required as well as voluntary steps taken for this purpose. Therefore, grouping of the items under different categories was necessary. While dividing these items into various groups, there was a risk of overlapping of certain items. Such items were placed under the group with which they were most directly related.

All the items included in the 'Index' were not of equal importance. Some items were apparently more important than others. The CAs were requested to evaluate each of these items independently and assign them points on the basis of their importance and usefulness in making sound investment and other decisions and in evaluating environmental performance of a company. They were requested to assign points to each individual item of information on following basis:

- if the item is unimportant;
- if the item is slightly important;
- if the item is moderately important;
- if the item is important, and
- if the item is very important.

Exhibit 6.1 shows the analysis of the responses. The last column in the Exhibit depicts the mean score for each item, calculated by averaging the points assigned by the CAs to a particular item. This 'mean score' is a measure of degree of importance that the CAs as a group assigned to that item. An information item is labelled as 'most important' if the mean score falls between 3.5-4.0. If the mean lies between 2.5 to 3.49, the item is considered as 'important'. If the mean is between 1.5 to 2.49, the item is 'moderately important'. Similarly, if the mean falls between 0.5 to 1.49, the item is regarded as 'slightly important'. Finally, a mean that falls below 0.49, is considered to be an indication that the related information item is unimportant.

Exhibit 6.1: Index of Environmental Disclosure

Number (Percent)

Items of Information	Points				
	0	1	2	3	4
1. Organisational overview					
(a) Broad environmental objectives	1 (1.0)	6 (6.0)	9 (8.9)	38 (37.6)	47 (46.5)
(b) Specific targets	2 (2.0)	9 (8.9)	26 (25.7)	34 (33.7)	30 (29.7)
(c) Management's environmental policy statement	1 (1.0)	5 (5.0)	16 (15.8)	40 (39.6)	39 (38.6)
(d) Environmental budget	2 (2.0)	13 (12.9)	19 (18.8)	29 (28.7)	38 (37.6)
(e) Environmental management system	–	10 (9.9)	11 (10.9)	39 (38.6)	41 (40.6)
(f) Environmental auditing	5 (5.0)	13 (12.9)	18 (17.8)	27 (26.7)	38 (37.6)
(g) Legal compliance	5 (5.0)	6 (6.0)	13 (12.9)	19 (18.8)	58 (57.4)
(h) Awards obtained for environmental protection	14 (13.9)	19 (18.8)	33 (32.7)	18 (17.8)	17 (16.8)
2. Prevention or repair of environmental damage					
(a) (e.g. reusing disposable material; waste management)	-	3 (3.0)	5 (5.0)	33 (31.7)	61 (60.4)
3. Aesthetic improvement					
(a) (e.g. Tree plantation)	1 (1.0)	5 (5.0)	19 (18.8)	26 (25.7)	50 (49.5)
4. Pollution control measures					
(a) Statutorily required	1 (1.0)	1 (1.0)	2 (2.0)	16 (15.8)	81 (80.2)
(b) Voluntary	2 (2.0)	4 (4.0)	10 (9.9)	31 (30.7)	54 (53.5)
5. Conservation of natural resources					
(a) Statutorily required	1 (1.0)	3 (3.0)	3 (3.0)	20 (19.8)	74 (73.3)
(b) Voluntary	2 (2.0)	5 (5.0)	9 (8.9)	33 (32.7)	52 (51.5)

6. Environmental accounting					
(a) Environmental costs	-	6 (6.0)	25 (24.7)	38 (37.6)	35 (34.7)
(b) Environmental liabilities and assets	1 (1.0)	9 (8.9)	27 (26.7)	37 (36.6)	27 (26.7)
(c) Significant accounting and reporting practices	4 (4.0)	5 (5.0)	20 (19.8)	43 (42.6)	29 (28.7)
(d) Extraordinary items	2 (2.0)	10 (9.9)	17 (16.8)	36 (35.6)	36 (35.6)
7. Project planning and management					
(a) Environmental management	2 (2.0)	13 (12.9)	27 (26.7)	42 (41.6)	17 (16.8)
(b) Environmental risk management	2 (2.0)	15 (14.8)	24 (23.7)	37 (36.6)	23 (22.7)
(c) Accident and emergency response	4 (4.0)	11 (10.9)	10 (9.9)	30 (29.7)	46 (45.6)
8. Social costs					
(a) Impact of organizational activities on environment	2 (2.0)	4 (4.0)	20 (19.8)	35 (34.6)	40 (39.6)
(b) Significant environmental risks not required to be disclosed legally	4 (4.0)	14 (13.8)	18 (17.8)	43 (42.6)	22 (21.7)
9. Any other	-	-	-	-	-

Contd...

Items of Information	Mean Score	Importance Category
1. Organisational overview		
(a) Broad environmental objectives	3.23	Important
(b) Specific targets	2.8	Important
(c) Management's environmental policy statement	3.1	Important
(d) Environmental budget	2.87	Important
(e) Environmental management system	3.1	Important
(f) Environmental auditing	2.79	Important
(g) Legal compliance	3.18	Important
(h) Awards obtained for environmental protection	2.05	Moderately Important
2. Prevention or repair of environmental damage		
(a) (e.g. reusing disposable material; waste management)	3.50	Most Important
3. Aesthetic improvement		
(a) (e.g. Tree plantation)	3.18	Important
4. Pollution control measures		
(a) Statutorily required	3.73	Most Important
(b) Voluntary	3.30	Important

5.	**Conservation of natural resources**		
(a)	Statutorily required	3.61	Most Important
(b)	Voluntary	3.27	Important
6.	**Environmental accounting**		
(a)	Environmental costs	2.95	Important
(b)	Environmental liabilities and assets	2.79	Important
(c)	Significant accounting and reporting practices	2.87	Important
(d)	Extraordinary items	2.93	Important
7.	**Project planning and management**		
(a)	Environmental management	2.58	Important
(b)	Environmental risk management	2.63	Important
(c)	Accident and emergency response	3.02	Important
8.	**Social costs**		
(a)	Impact of organizational activities on environment	3.06	Important
(b)	Significant environmental risks not required to be disclosed legally	2.64	Important
9.	**Any other**	0	-

Note: 0–Unimportant; 1–Slightly Important; 2–Moderately Important; 3–Important; 4– Most Important

The Exhibit shows that the CAs indicated preference for a broad range of information items relating to corporate environmental reporting. Three items of environmental information namely, 'prevention or repair of environmental damage', 'statutorily required pollution control measures' and 'statutorily required measures taken for conservation of natural resources' were considered as 'most important' by the qualified accountants. 19 items were regarded as 'important' by them. Only one item, which had been considered to be only of moderate importance, was 'awards obtained for environmental protection'. [2] It is also important to note that no CA considered information on 'environmental costs' as 'unimportant'. This shows increasing awareness among CAs about environmental accounting. Another important feature about the responses was that the respondents had not added 'any other item(s)' to the 'Index', although requested to do so, if they so desired. This shows comprehensiveness of the 'Index' used in the study. Exhibit 6.2 shows frequency of these items under various importance categories.

6.1.2 Analysis of Annual Reports: For examining

corporate environmental disclosure practices in India, annual reports of 91 companies [3] were analysed for three consecutive years 1999-2000, 2000-2001 and 2001-2002. The annual reports were used for this purpose because they are considered to be the most widely used and popular mode of disclosing any relevant information. [4] These reports are authenticated by external qualified auditors.

Exhibit 6.2: Frequency of Items According to Importance

Importance Category	Range	Number of Items
(a) Most Important	3.5-4.0	3
(b) Important	2.5-3.49	19
(c) Moderately Important	1.5-2.49	1
(d) Slightly Important	0.5-1.49	-
(e) Unimportant	0-0.49	-
Total	-	**23**

Moreover, all the public limited companies are required to produce these reports regularly due to legal compulsion. Since these reports are produced by all the companies, it makes inter-company comparisons relatively easy. The annual report is also considered to be a major medium for a company to promote itself and the inclusion of other information with the financial information in this report, may indicate relative importance of that information. [5] An analysis of green reporting practices over years would help in checking the consistency in these disclosures.

For examining environmental disclosure practices of the sample companies, various worksheets were prepared for every company for all the three years, detailing the environmental information reported by it. After completing worksheets for all the sample companies, for each item of information, the number of companies disclosing it was computed and divided by the total number of companies in the sample. In this way, disclosure percentage of each item of environmental information was derived. An attempt was also made to examine the place of this disclosure in the annual report, its type and length. Thus, both quantity and quality of environmental disclosures were checked in the study.

To evaluate disclosure practices followed by the individual companies, environmental disclosure score (EDS) percentages were calculated for each of them. These were obtained by using two methods:

1. by giving equal weights to all the items of information;
2. by giving differential (unequal) weights according to the perceptions of the CAs. [6]

The weights received by a company for the information disclosed in the annual report were summed up and divided by the maximum score that could be obtained by it. The quotient (calculated as percentage) was termed as 'EDS percentage'. In this way, for every company, the following calculations were made:

- **Maximum EDS:** This measured the maximum amount of environmental information, which could be reported in the annual report by a company. This score was common for all the companies. Exhibit 6.3 summarizes the weights given to each of the items in the 'Index'. It can be seen in the Exhibit that by assigning equal weights to all the items, the 'maximum score' was 23. By assigning differential weights as given by the CAs, this score was 69.18.
- **Actual EDS:** It measured the amount of environmental information actually found in the annual report of a company. Separate scores were calculated by using both equal and unequal weights.
- **EDS Percentage:** It depicted the amount of environmental information reported in the annual report of a company as a percentage of maximum amount of information that could have been reported. Thus, for all the companies, two EDS percentages were calculated for each of the three years—one by using equal weights and another by assigning differential weights.

6.1.3 Multivariate Analysis: Environmental disclosure practices are affected by number of company and industry related factors (e.g., large size private sector companies having foreign links might be disclosing more environmental information). All the factors taken together may give a better explanation to the variation in these practices. Therefore, an attempt was made in the study to examine the relationship between environmental disclosure practices and all the selected variables with the help of multiple regression technique.

Exhibit 6.3: Weights Given to Each of the Items in the Index

Items of Information	Importance Category	Weights	
		Equal	Unequal (Mean Score)
1. Organisational Overview			
(a) Broad environmental objectives	Important	1	3.23
(b) Specific targets	Important	1	2.8
(c) Management's environmental policy statement	Important	1	3.1
(d) Environmental budget	Important	1	2.87
(e) Environmental management system	Important	1	3.1
(f) Environmental auditing	Important	1	2.79
(g) Legal compliance	Important	1	3.18
(h) Awards obtained for environmental protection	Moderately Important	1	2.05
2. Prevention or Repair of Environmental Damage			
(a) (e.g. reusing disposable material; waste management)	Most Important	1	3.50
3. Aesthetic Improvement			

Contd...

(a) (e.g. Tree plantation)	Important	1	3.18
4. Pollution Control Measures			
(a) Statutorily required	Most Important	1	3.73
(b) Voluntary	Important	1	3.30
5. Conservation of Natural Resources			
(a) Statutorily required	Most Important	1	3.61
(b) Voluntary	Important	1	3.27
6. Environmental Accounting			
(a) Environmental costs	Important	1	2.95
(b) Environmental liabilities and assets	Important	1	2.79
(c) Significant accounting and reporting practices	Important	1	2.87
(d) Extraordinary items	Important	1	2.93
7. Project Planning and Management			
(a) Environmental management	Important	1	2.58
(b) Environmental risk management	Important	1	2.63
(c) Accident and emergency response	Important	1	3.02
8. Social Costs			
(a) Impact of organizational activities on environment	Important	1	3.06
(b) Significant environmental risks not required to be disclosed legally	Important	1	2.64
Total	–	23	69.18

Main Hypothesis: The main hypothesis tested in the study was that the extent of environmental disclosure is a function of various industry- and company-related variables. Specifically, following hypotheses were tested:

- H_1 = Environmental information disclosure by public

sector companies is significantly higher than private sector companies.

- H_2 = Most polluting industries disclose significantly more information on environment related issues than less polluting industries.
- H_3 = Extent of environmental reporting in various industries is different.
- H_4 = Companies having foreign association disclose more information on environmental issues than domestic companies.
- H_5 = Environmental information disclosure by group and non-group companies is significantly different.
- H_6 = The extent of environmental disclosure is positively related to size of the company.
- H_7 = An increase in profitability influences reporting on environment positively.
- H_8 = Environmental reporting is positively related to export performance of a company.
- H_9 = High debt-equity ratio leads to increased reporting on environmental issues.
- H_{10} = The extent of environmental disclosure is positively associated with environmental performance of the company.

No study seems to have been made in India using this type of analysis. However, a few works conducted abroad used multivariate analysis to test this kind of relationship. An attempt was made to compare observations made in this Chapter with those of similar other research works. As observed earlier, the analysis in this Chapter combines primary information obtained from the companies with the help of the questionnaire and the secondary information collected from the annual reports of the companies. Therefore, the analysis is based on the 91 companies, for which primary as well as secondary data were available.

6.2 Analysis of Environmental Disclosures Practices

This Section attempts to present an analysis of

environmental disclosure practices followed by the selected companies. The Section is divided into four sub-sections. Sub-section 1 evaluates extent of environmental information disclosure in published annual reports of the companies. Sub-section 2 presents the type of information reported. In sub-section 3, the place where these details are disclosed in the annual report is analysed. Finally, sub-section 4 examines the length of the disclosure.

6.2.1 Extent of Disclosure: As mentioned before, to examine the extent of environmental information disclosed by a company, the 'Index' constructed in primary data analysis was used. The 'Index' consisted of 23 items of information. While three items were considered 'most important' by the qualified accountants, rest of them (excluding one) were regarded as 'important' (Exhibit 6.3).

Exhibit 6.4: Extent of Disclosure

Number (Percent)

Items of Information	Companies disclosing information		
	1999-2000	**2000-2001**	**2001-2002**
I. Organisational Overview			
(a) Broad environmental objectives	19 (20.9)	22 (24.2)	24 (26.4)
(b) Specific targets	8 (8.8)	4 (4.4)	7 (7.7)
(c) Management's environmental policy statement	44 (48.4)	43 (47.3)	48 (52.7)
(d) Environmental budget	3 (3.3)	4 (4.4)	4 (4.4)
(e) Environmental management system (EMS)	17 (18.7)	18 (19.8)	31 (34.1)
(f) Environmental audit	18 (19.8)	20 (22)	20 (22)
(g) Legal compliance	45 (49.5)	41 (45.1)	37 (40.7)
(h) Awards obtained for environmental protection	26 (28.6)	30 (33)	28 (30.8)

Contd...

2.	**Prevention or Repair of Environmental Damage (e.g. reusing disposable material)**	27 (29.7)	23 (25.3)	23 (25.3)
3.	**Aesthetic Improvement (e.g. tree plantation)**	24 (26.4)	25 (27.5)	19 (20.9)
4.	**Pollution Control Measures**			
(a)	Statutorily required	43 (47.3)	41 (45.1)	41 (45.1)
(b)	Voluntary	19 (20.9)	19 (20.9)	19 (20.9)
5.	**Conservation of National Resources**			
(a)	Statutorily required	91 (100)	91 (100)	91 (100)
(b)	Voluntary	23 (25.3)	21 (23.1)	25 (27.5)
6.	**Environmental Accounting Information**			
(a)	Environmental Costs	14 (15.4)	9 (9.9)	7 (7.7)
(b)	Environmental assets and liabilities	6 (6.6)	3 (3.3)	1 (1.1)
(c)	Significant accounting and reporting policies	-	1 (1.1)	-
(d)	Extraordinary items	-	-	1 (1.1)
7.	**Project Planning and Management**			
(a)	Environmental management	5 (5.5)	3 (3.3)	6 (6.6)
(b)	Environmental risk management	6 (6.6)	7 (7.7)	6 (6.6)
(c)	Accident and emergency response	8 (8.8)	12 (13.2)	11 (12.1)
8.	**Social Costs**			
(a)	Impact of organisation's activities on environment	3 (3.3)	2 (2.2)	4 (4.4)
(b)	Significant environmental risks not required to be disclosed legally	3 (3.3)	2 (2.2)	4 (4.4)
9.	**Any other**	5 (5.5)	5 (5.5)	2 (2.2)

Exhibit 6.4 displays the number and the percentage of companies disclosing different items of environmental information over the period of three years, i.e., 1999-2000, 2000-2001 and 2001-2002.

The Exhibit shows that about one-fifth of the companies disclosed the item 'broad environmental objectives' in the year 1999-2000 and this percentage increased to about 26 percent in 2001-2002. Generally, these objectives were given in two to three lines, either on the front page of the annual report or in the company's mission or vision statement. A few companies (mainly some large public sector companies) published their environmental objectives separately in the annual report. Item 'specific targets' was disclosed by less than nine percent of the companies in all the three years. [7] More than one-half of the companies reported their 'environmental policies' in 2001-02. A large majority of these companies disclosed these policies only to show their commitment towards environmental goals. But most of the times, this information was given in a few lines only in the Director's Report. Less than five percent of the companies published some figures on 'environmental budget', though this item was considered as 'important' by the CAs. Even the companies giving this information reported in one to two lines the likely amount to be spent for environmental cause.

About 19 percent of the companies mentioned 'environmental management systems (EMS)' (its introduction/ continuation) in 1999-2000. The percentage increased significantly to 35 percent in 2001-2002. An important reason for this increase could be a significant growth in the number of companies obtaining ISO 14001 certificates these days. It is to be noted that one of the requirements for getting this certificate is establishment of an efficient system of EMS. About 20-22 percent of the companies mentioned about environment and/or safety audits in the annual reports over the years. A relatively significant percentage (about one-half) of the companies talked about 'legal compliance' in 1999-2000; though the percentage decreased slightly to 41 percent in 2001-2002. It seems that these days, there is a shift in emphasis from reporting legal compliance to disclosure of initiatives undertaken in this area. The only item, which had been considered as of 'moderate importance' by the CAs was 'awards obtained for environmental protection'. About one-third of the companies reported these awards in different years. It is obvious that a company is and should be proud of these awards as it displays its commitment towards social goals and is evidence that the company is a good corporate citizen.

The item 'prevention or repair of environmental damage' was considered as the 'most important' by the CAs. But only three-tenths of the companies disclosed this item in 1999-2000; the percentage decreased to about 25 percent in 2001-2002. Almost same trend was visible in case of the next item 'aesthetic improvement'. Another item regarded as 'most important' by the CAs was statutorily required measures taken for pollution control. About 45-47 percent of the companies reported information on this item over the years. Only about one-fifth of the companies disclosed voluntary steps taken to combat pollution. It appeared that a majority of the companies did not take initiatives in this regard and that was the reason for not disclosing this item. Reporting on 'conservation of natural resources' was also considered as 'most important' by the CAs.

It is to be noted that the information on conservation of energy is legally required to be disclosed by all the public limited companies in the annual report under Section 217(1)(e) of the Companies Act, read with the Companies (Disclosure of Particulars in the Report of Board of Directors) Rules, 1988. Companies are required to disclose this information every year in the annexure to Director's Report. Therefore, all the companies provided detailed information on this item every year (steps taken for conserving energy as well as benefits received therefrom). [8] But, again only a small proportion of the companies voluntarily disclosed information on conservation of other natural resources. This makes a strong case for making environmental disclosures mandatory in India. Frost (2001) found a significant improvement in overall reporting by Australian companies on their environmental performance after introduction of mandatory reporting in 1998. [9] However, Larrinaga (1999) found otherwise in a study on evaluation of mandatory environmental accounts in Spain. He did not find any improvement in environmental reporting after introduction of mandatory environmental reporting requirement. [10]

Regarding the category 'environmental accounting information', though the accountants considered all the items under this category as 'important', very few companies disclosed this information in the annual report. The only item disclosed by some of the companies in this category was 'environmental costs'. However, most of these companies only gave figures of significant amount spent for environmental purposes in a particular year in the Director's Report. Only a few concerns separately recognised this amount in the Profit & Loss Account as 'green belt development or horticulture expenses'. No company (except one) reported any 'significant accounting or reporting policies' or 'extraordinary items' in the annual report. This shows that in India, quantitative/financial reporting on environmental issues is still at infancy stage. K. Jones (2000) in a survey of annual/financial reports of 88 EU companies also found that very few companies disclosed quantified information on environmental issues (e.g., environmental expenses and capitalised environmental costs). [11]

Before starting a project, it is very important to take care of its impact on the environment. In many cases environmental impact assessment is legally compulsory to obtain various licences. This item was reported by less than one-tenth of the companies over the years. Item 'accident and emergency response' was disclosed by 9-13 percent of the companies, mostly under the heading 'environment, health and safety issues' or 'number of accident free days in a year'. Disclosure of information on 'social costs' was again negligible. Some companies disclosed 'other items' also, like 'training for environmental protection' or 'celebration of environment day' or 'innovations in the field'.

In nutshell, though all the companies disclosed statutorily required information on energy conservation, the percentage of the companies disclosing most of the other items was still low. Some of the companies voluntarily disclosed information on environmental policies, objectives, legal compliance, and measures taken for pollution control. However, only a negligible number reported environmental accounting information, information on project planning and management, and social costs.

6.2.2 Type of Information Disclosed: As noted above, a company can disclose qualitative information on environment (e.g. information on environmental objectives, policies or audit). In addition quantitative and/or financial information can also be disclosed by a company. Exhibit 6.5 presents the type of environmental information disclosed by the companies over three years.

Exhibit 6.5: Type of Disclosure

Number (Percent)

Type of Disclosure	Companies disclosing information		
	1999-2000	2000-01	2001-02
(a) Quantitative only	-	-	-
(b) Qualitative only	76 (83.5)	78 (85.7)	82 (90.1)
(c) Both	15 (16.5)	13 (14.3)	9 (9.9)
Total	91 (100)	91 (100)	91 (100)

It was observed that the common practice followed by the companies regarding environmental disclosure was to offer descriptive information in the annual report. This trend increased overtime. Only a few companies provided quantitative as well as qualitative information on environment. Moneva and Llena (2000) in a study of 70 large companies operating in Spain also found that environmental reporting was mainly narrative in nature, with a very limited use of strictly financial data.

However, they found a continuous increase in quantitative and financial reporting during the period 1992-94. [12] O'Dwyer (2003) in a study of 83 listed Irish companies observed that only 12 percent of the companies made financial disclosures, mainly to comply with new financial reporting requirements with respect to provisions for environmental liabilities. [13] The Exhibit shows that in 2001-02, about 90 percent of the companies disclosed only qualitative information in the annual reports. Even the companies which disclosed financial information on environmental issues did not provide any item-wise break up of expenditure or its accounting treatment in these reports. Moreover, there was no consistency in this kind of reporting. However, some companies gave elaborate information through charts and tables on pollution levels or emission of pollutants. Thus, majority of the companies disclosed only qualitative/descriptive information on the environment in the annual report.

6.2.3 Place of Environmental Disclosure: The environmental information can be disclosed at various places in the annual report like in the main body of financial statements, Director's Report or in a separate section in this report. Exhibit 6.6 depicts practices of the sample companies regarding location of environmental disclosures in the annual report.

The Exhibit shows that most of the companies made environmental disclosures in the Director's Report. This trend has increased overtime. Roy (2000) also observed that the common practice followed by the Indian companies in both public and private sector over the years regarding environmental disclosure was to offer descriptive information and itemize the same in the Director's Report. Only a few companies disclosed information through notes, schedules and a separate social account. [14] Jaffer et al. (2002) also found most of the environmental information in the Review of operations section and in Chairman's statements. [15] As compared to this, Jones (2000) in a study of 88 European companies for environmental disclosures found that out of 13 sections identified in the annual reports, three-fourths of the companies provided environmental information in the 'environmental section'. [16]

Exhibit 6.6: Place of Disclosure

Number (Percent)

Location in annual report	Companies making Disclosure		
	1999-2000	**2000-2001**	**2001-2002**
(a) Main body of financial statements	-	1 (1.1)	-
(b) Footnotes to the financial statements	1 (1.1)	-	-
(c) Director's report/management discussion and analysis	53 (58.2)	55 (60.4)	62 (68.1)
(d) Annexure to Director's report	91 (100)	91 (100)	91 (100)
(e) Separate annexure to financial statements	-	-	1 (1.1)
(f) Separate environmental section	5 (5.5)	5 (5.5)	3 (3.3)
(g) Any other	14 (15.4)	12 (13.2)	11 (12.1)

All the companies reported the item 'conservation of energy' in 'Annexure to Director's Report' as it is compulsory under the Companies Act. However, some companies provided this information in separate environmental section in the annual report. The purpose might be to provide detailed information on the issue as well as to highlight organisation's commitment towards social goals. No company (except one) disclosed environmental information in the main financial statements or its footnotes or in separate annexure to it. Some companies used other locations also for environmental disclosure like first page of the annual report, Chairman's message and booklets attached with the annual reports. Thus, majority of the companies disclosed voluntary information on environment in the Director's Report, in addition to statutorily required information in its annexure.

6.2.4 Length of Voluntary Environmental Disclosure: It is evident from the previous sub-sections that a large majority of the sample companies reported only qualitative/descriptive environmental information in the annual reports. Most of them gave only statutorily required information on energy conservation. However, a significant number of the concerns also provided these details voluntarily (e.g., information on environmental policies, audits or legal compliance was provided by some of them). In this sub-section, an attempt has been made to check the length of these voluntary disclosures. [17] Exhibit 6.7 gives an analysis of the length of environmental disclosure made by the companies.

Exhibit 6.7: Length of Voluntary Environmental Disclosure

Number (Percent)

Number of Pages	**Companies disclosing**		
	1999-2000	**2000-2001**	**2001-2002**
(a) Zero page	36 (39.6)	32 (35.2)	25 (27.5)
(b) Less than a half page	31 (34.1)	34 (37.4)	38 (41.8)
(c) Half to one page	13 (14.3)	14 (15.4)	15 (16.5)
(d) One to two pages	4 (4.4)	5 (5.5)	8 (8.8)
(e) More than two pages	7 (7.7)	6 (6.6)	5 (5.5)
Total	**91 (100)**	**91 (100)**	**91 (100)**

As shown in the above Exhibit, about four-tenths of the companies had not disclosed any voluntary information in 1999-2000. However, this percentage decreased to about 28 percent in 2001-02. In other words, the percentage of the companies reporting voluntarily on the issue increased over this period. About 42 percent of the companies made disclosure of 'less than a half page' in the annual reports in 2001-2002. The results revealed that though a significant percentage of the companies made voluntary disclosures on the issue, in most of the cases the length of this disclosure ranged from two to three lines to a half page. However, some large companies gave this information in detail in the separate environmental sections. Findings of PIRC (1996) survey showed that the amount of environmental information provided by FTSE 350 companies ranged from one sentence to externally verified statements running over 30 pages. [18] But, Roy in a study of annual reports of 55 Indian companies for the years 1987, 1991 and 1996, found that on all environmental issues (except energy conservation), the extent of disclosure did not exceed even one-fourth page in most of the cases. [19]

6.3 Environmental Disclosure Scores (EDS)

This Section presents an analysis of the EDS percentages obtained by the sample companies over the years. The analysis would help in examining the extent of environmental disclosures made by these companies. The Section is divided into two sub-sections. While sub-section 1 gives EDS percentages calculated by using equal weights; sub-section 2 shows these percentages derived by using differential weights.

6.3.1 EDS Percentages (Using Equal Weights): Exhibit 6.8 depicts the division of the companies on the basis of their EDS percentages calculated (for each of the three years) by using equal weights.

Exhibit 6.8: EDS Percentages (Using Equal Weights)

Number (Percent)

EDS (as percent)	Companies		
	1999-2000	**2000-2001**	**2001-2002**
(a) Less than 25	57 (62.64)	60 (65.93)	62 (68.13)
(b) 25-50	24 (26.37)	23 (25.27)	17 (18.68)
(c) 50-75	8 (8.79)	7 (7.69)	12 (13.19)
(d) More than 75	2 (2.2)	1 (1.09)	–
Total	**91 (100)**	**91 (100)**	**91 (100)**

The Exhibit shows that majority of the companies in all the three years were in 'less than 25 percent' EDS group. This group mainly comprises of companies disclosing only the statutorily required information on energy conservation. This percentage has increased over-time. About one-fourth of the companies were in '25-50 percent' EDS group in 1999-2000; but this percentage decreased over years. There was a slight increase in the percentage of companies in '50-75 percent' group. The results indicate lack of consistency in reporting by some of the companies over years.

6.3.2 EDS Percentages (Using Unequal Weights): Exhibit 6.9 presents the division of the companies on the basis of EDS percentages obtained by them during each of the three years using differential weights.

Exhibit 6.9: EDS Percentages (Using Unequal Weights)

Number (Percent)

EDS (as percent)	**Companies**		
	1999-2000	**2000-2001**	**2001-2002**
(a) Less than 25	58 (63.74)	59 (64.84)	61 (67.03)
(b) 25-50	23 (25.27)	21 (23.08)	16 (17.58)
(c) 50-75	8 (8.79)	9 (9.89)	14 (15.38)
(d) More than 75	2 (2.2)	2 (2.2)	–
Total	**91 (100)**	**91 (100)**	**91 (100)**

The Exhibit reveals almost the same trend as shown by the Exhibits 6.8. Majority of the companies in all the three years were from 'less than 25 percent' EDS group, followed by the companies in '25-50 percent' group. While the percentage of companies in 'less than 25 percent' and '50-75 percent' groups increased overtime, it decreased in '25-50 percent' group. The results indicate that majority of the companies in India do not make adequate disclosures on environmental issues. The quality as well as quantity of these disclosures is quite limited. Moreover, there is no consistency in this reporting.

6.4 Regression Analysis

This Section introduces basic regression analysis and the

measures used to evaluate the estimated regression equations. While sub-section 1 gives an introduction to basics of regression analysis, sub-section 2 explains building up of model for the present study.

6.4.1 Regression Analysis: An Introduction: Regression analysis is "concerned with the study of the dependence of a variable, the dependent variable, on one or more other variables, the explanatory variables, with a view to estimating and/or predicting the (population) mean or average value of the former in terms of the known or fixed (in repeated sampling) values of the latter." [20] In a regression equation model, the casual relationship between a dependent variable and one or more independent or explanatory variables is expressed as a linear function of the parameters. [21]

The regression equation can be of two types–simple regression and multiple regression equations. Simple or two variable regression analysis is a study of the dependence of a variable only on a single explanatory variable. Multiple regression, on the other hand, helps in observing the relative importance of different independent variables in providing changes in the dependent variable. In the present study, multiple regression analysis was used.

The general form of multiple regression equation is:

$Y = \square + \square_1 X_1 + \beta_2 X_2 + \ldots\ldots + \beta_n X_n + e$ where

Y is the dependent variable or regressand;

$X_1, X_2 \ldots\ldots\ldots X_n$ are independent variables or regressors;

$\square$ is the intercept;

$\square_1, \square_2 \ldots\ldots \square_n$ are the parameters or the regression coefficients and e is the error term which is $N(0, \sigma^2)$.

The regression coefficients tell about the change in the dependent variable (Y) for each unit change in the independent variable (say X_1), when the other independent variables (say $X_2 \ldots\ldots\ldots X_n$) are maintained at a given level. The method of ordinary least squares (OLS) is used to construct the regression function. The estimators of the regression parameters obtained by the method of least squares, with some assumptions [22], are optimum.

In regression analysis, the dependent variable or regressand is frequently influenced not only by ratio scale variables (like income, output, prices, costs, height or temperature), but also by variables that are essentially qualitative or nominal scale in nature (such as sex, race, colour, religion or nationality). Since such variables usually indicate the presence or absence of a quality or an attribute, such as male or female, black or white, they are essentially nominal scale variables. One way to 'quantify' such attributes is by constructing an artificial variable that takes on values of '1' or '0'; '1' indicates the presence (or possession) of that attribute and '0' indicating the absence of that attribute. For example, 1 may indicate that a company is a public sector company and 0 that it is a private sector company. Variables that assume such 0 and 1 values are called dummy variables. [23] Such variables are thus, essentially a device to classify data into mutually exclusive categories such as male or female. In the present study also, dummy variables were used to incorporate qualitative variables like sector, foreign association and industry into the regression equation.

The estimated regression equations were evaluated on the basis of the signs and the size of the estimated parameters, coefficient of correlation among the individual variables, R^2 or adjusted R^2, F test and t test.

The signs (plus or minus) of the estimated parameters tell about the type of relationship (positive or negative) between an independent variable and the dependent variable. Size of the estimated parameters indicates the magnitude of each unit change in an independent variable and the resultant variation in the dependent variable when the effect of other independent variables is held constant. In case of dummy variables, the regression coefficient indicates by how much the mean values of two categories differ. If there are differences in the response of the regressand to the variation in the qualitative variables in the various sub-groups, they will be reflected in the intercepts or slope coefficients or both of the various subgroup regressions. [24]

The coefficient of correlation among the explanatory variables helps in finding out the degree of multicollinearity. R^2, multiple coefficient of determination, is a measure of the proportion of the variance in the dependent variable, which is explained by the independent variables in the regression equation. Adjusted R^2, known as the 'corrected coefficient of determination' has been used to measure the 'goodness of fit'. This measure takes into account the number of explanatory variables in relation to the number of observations. The purpose of adjusted R^2 is to facilitate comparison of several regression equations that may vary with respect to the number of explanatory variables and the number of observations.

Statistical measures such as 't test' and 'F test' were employed to determine the significance of the percentage of variation in the dependent variable explained by the successive addition of the exogenous variables and to test the significance of degree of association and inter-relationship between different variables. F-test was applied to test whether the relationship is significant or not. It is a measure of testing significance of R^2 or R^2. The t-test was used for testing the significance of individual parameters. In the present study, the reliability of the results were determined with the help of most of these well accepted measures. Thus, the study makes use of all the three – a priori, statistical and econometric criteria.

6.4.2 Building the Model: For empirically testing the impact of selected variables on the environmental disclosure practices of the companies, two types of multiple regression equations were used. These equations were named Type I and Type II regression equations. In the absence of contrary information, linear relationship was presumed among the variables in each of the equations. The signs of different parameters were based on theoretical considerations.

Identification of Variables:

(A) Dependent Variable: As mentioned before, this

Section tries to examine influence of the selected variables on environmental disclosure practices of the companies. In the previous Section, environmental disclosure score (EDS) percentages were calculated for every company for each of the three years (1999-2000, 2000-01 and 2001-02) using both equal and unequal weights (weights given to various items of environmental information by the CAs). The average of EDS for 3 years was calculated. Two averages—one for EDS calculated by using equal weights and another for EDS with unequal weights, were taken as dependent variables in two separate regression equations.

(B) **Independent Variables:** The independent variables were identified on the basis of their theoretical relevance in explaining environmental disclosure practices of the companies. The selection was guided by the theoretical considerations as well as the results of empirical evidences found in primary data analysis. The selected variables and their likely impact on environmental reporting have been described briefly in this sub-section. The hypotheses given in Section 6.1 were also based on the theoretical considerations explained here. The selected independent variables were:

1. **Sector:** The sample comprises of both public and private sector companies. The public sector companies are expected to disclose more environmental information as they are subject to wider exposure through evaluation in Parliament and/or legislature. Moreover, these companies generally operate with social and environmental objectives and also have favourable attitude towards disclosure. They have to follow stringent rules and have to be more transparent in their activities.
2. **Nature of Industry (According to product produced):** The extent of environmental disclosure in annual reports may differ from industry to industry. It is due to a number of reasons. Firstly, the fear that the disclosure of information could possibly aid the company's competitors varies among industries and thereby, may lead to industry

difference in the extent of disclosure. Generally, in the industries where competition is perceived to be keen, the fear that disclosure of environmental information will adversely affect the competitive position of the company may be there. Hence, the temptation may exist for companies within that industry to conceal certain sensitive information from their competitors. Secondly, to maintain a healthy relationship between the company and investing community, each company within an industry may try to ensure that its annual report is as informative as those of its rivals. Consequently, this may cause industry difference in environmental disclosure.

3. **Nature of Industry (According to Pollution generated):** Though all manufacturing companies are polluting in nature, Environment (Protection) Rules, 1992 give a list of 'most polluting industries'. These industries have to submit an 'Environmental Statement' in a prescribed format to relevant State Pollution Control Boards (SCPB) every year. It is obvious that they take various steps to prevent and control pollution and to conserve natural resources because they have to report to the regulatory authorities. Since these companies spend significant amounts on environment, they are expected to disclose more environmental information as compared to less polluting companies.
4. **Foreign Association:** It has already been discussed that environmental protection rules are quite stringent in some countries like USA, UK and Australia. Most of these countries also have compulsory disclosure requirements to report this information in or outside financial statements. Moreover, emphasis is also being placed on voluntary disclosure of environmental information in these countries. As a result, companies operating in these countries tend to disclose a lot of environmental information in published annual reports. [25] It can also be assumed that companies operating in India, which have foreign association will also

disclose more environmental information than the domestic companies. By foreign association, it is meant here that the company is a subsidiary of a company incorporated outside India or foreign promoters have a share of more than 50 percent in the company.

5. **Control by large business houses:** Companies which are associated with large business houses like Birlas and Tatas, are expected to report more on environment than non-group companies.
6. **Size of the company:** There are several possible reasons for expecting a positive relationship between size of a company and the extent of environmental disclosure in the annual report. Firstly, the accumulation and dissemination of information is a costly affair. Smaller companies may not have necessary resources to gather the huge amount of information and to disseminate it to the public. Secondly, in large companies such information is generally collected for internal reporting and subsequently used in managerial decision-making. [26] They already have this information and can easily report relevant information to external stakeholders also. Thirdly, large companies are generally subject to more public expectation and watch by government authorities. Moreover, these companies usually have more shareholders (individual and institutional) to satisfy with information. Such companies believe that disclosure of environmental information will create confidence among stakeholders and will minimize the public criticism and pressures from government. Therefore, large companies are expected to disclose more environmental information than small companies. Total paid-up capital of the company as on 31st March 2002 was used as a measure of 'size' in the study because it is more stable as compared to other measures of size like turnover, number of shareholders or net tangible assets.
7. **Profitability:** In the present study, profitability has been defined as the ratio of net profits to net sales. It is the

overall measure of management's ability to turn each rupee of sales into profit. If the profitability is high, the company will have additional capacity to absorb rising costs. The position of the company thereby, would become stable and strong to face the competition in the market. If the earnings margin of a company is more than that of its industry, the public will have greater confidence in the company. The company would than prefer to disclose more information to better inform the public about social/environmental activities undertaken by it. If the earnings margin is low, then the company cannot be expected to spend more on social obligations and will also hesitate in disclosing it. Thus, a company with higher earnings margin is expected to disclose detailed information on environmental activities than companies with low profitability.

8. **Debt-equity Ratio:** Many institutional investors (particularly foreign institutional investors like World Bank) require information on environmental performance of a company. A company taking loan from these institutions (for new or existing projects) has to submit a due diligence certificate disclosing impact of its activities on the environment, significant environmental risks and steps taken to protect and conserve environment. Therefore, the companies with high debt-equity ratio are expected to disclose more environmental information than the companies with lower debt -equity ratio.
9. **Exports:** A company entering into export market has to compete with global players. Most of the multinational corporations incorporated in developed countries follow stringent environmental rules, obtain ISO 14000 certificates and claim to produce eco-friendly products. Moreover, due to UNO efforts and popularity of green movement; environmental care, green initiative and sustainable development have become a buzzword in the global market. Obviously, a company exporting its product

to other countries has to take care of international developments. An attempt has been made to see association between export performance and environmental disclosure practices of the companies. Export as percentage of sales has been taken as one of the independent variable in the multiple regression equation used in the study.

10. **Environmental Performance (EP):** In the present study, an attempt was made to relate environmental reporting with EP of the company. In general, a company which performs better on environment is expected to report more environmental information. Since information on EP of a company was not available in the annual report and in various databases, it was measured on the basis of primary data obtained on this aspect from the companies. It was measured indirectly, on the basis of information obtained on 3 broad issues:

(a) Existence of environmental policies and systems: This includes existence of:

- Environmental policy
- Separate environmental department
- Environmental management system
- ISO 14001 certificate

(b) Preparation of Environmental Accounts and Statements:

- Accounts for external reporting or internal decision-making
- Environmental statements (statutorily required or voluntary)
- Audit of environmental statements
- Recognition of social costs, its valuation and consideration in decision-making

(c) Operational aspects like training to employees, internal reward system, awards obtained for better environmental performance and voluntary steps taken for environmental protection.

Weights were assigned to all these items in two different ways. The weights were:

i) **Equal Weights**: Here, equal weights were given to all the items. Weight '1' was given if a particular system was in existence and weight '0' was assigned if the system was non-existent. For example, if a company had an operating EMS or had plans to introduce it, score '1' was given, but if it did not have any plans to introduce EMS in near future, '0' weight was given. Similarly, for existence of environmental policy, environmental department or reporting systems weights were assigned.

ii) **Unequal Weights**: Under this system, differential weights were given to various items on the basis of theoretical considerations, the perceived importance of an item as indicated by observations made while collecting primary data and perception of the researcher. For example, regarding the category 'preparation of environmental reports and statements', the maximum weight (i.e., 2) was given if a company prepared these accounts for external reporting purposes. If these accounts were used for internal decision-making, score '1' was given as it was a common practice. If these accounts were not at all prepared, '0' weight was given. Similarly it was observed that the companies generally prepared environmental statements for legal purposes only. But some companies published voluntary statements also. Such a practice was generally followed if the company took initiatives to protect and conserve environment and had some information to report to the public. Hence, if a company prepared voluntary environmental statements, it was assigned weight '2'; if these statements were prepared only to comply with some legal provision, weight '1' was given. Furthermore, an additional weight '2' was given if the statements prepared by a company were audited by external independent auditor. But where these were audited by internal auditor, score '1' was given. Thus, if a company published externally verified environmental statements voluntarily, it got '4' points. If it produced only statutorily required externally verified information or an internally audited voluntary statement, weight '3' was given.

In the same way, differential weights were assigned to each item within a sub-classification (e.g. recognition of social costs). However, separate differential weights were not given to various sub-classifications within a broad category (e.g., to environmental policy, environmental department, EMS and ISO 14000 certificate) or to various broad categories. These items along with their weights are shown in Exhibit 6.10.

In measuring EP in this way, some limitations were there:

- Availability of limited information because of either reluctance on the part of respondents or fear of exposure to public.
- Overlapping of some items—e.g. environmental management system is also a part of ISO 14001 certificates. Similarly environmental audit can also be included in it. Items were put under different categories on the basis of perception of the researcher.
- Subjectivity in assigning weights to various parameters. Weights were assigned on the basis of perception of the researcher as well as theoretical considerations (wherever available). Though subjective, assignment of weights was necessary to convert these qualitative parameters into quantitative items.

Though measurement of EP was subject to above limitations, it was felt that in the absence of any direct way to measure it, this method would give a broad idea about EP of different companies. It was assumed that a company performing better on environment provides detailed information than poor environmental performers to publicize its environmental activities and to show that it is a good corporate citizen. [27]

As mentioned above, to examine the impact of nature of industry on environmental reporting, two classification systems were used. Firstly, the industries were categorized on the basis of pollution generated.

Exhibit 6.10: Measurement of Environmental Performance

Categories	Equal Weights	Unequal Weights
1. **Existence of Policies and Systems**		
(a) Environmental Policy		
• Yes	1	2
• No	0	0
(b) Separate Environmental Department		
• Yes	1	2
• No	0	1
(c) Environmental Management System		
• Already in existence	1	2
• Planned in near future	1	1
• No plans for it	0	0
(d) ISO 14001 certificate		
• Already obtained	1	2
• in the process of obtaining	1	1
• not yet obtained	0	0
2. **Preparation of Environmental Accounts and Statements**		
(a) Environmental Accounts		
• For External reporting		
♦ Yes	1	2
♦ No	0	0
• For Internal decision-making		
♦ Yes	1	1
♦ No	0	0
(b) Environmental Statements		
• Statutorily Required		
♦ Yes	1	1
♦ No	0	0
• Voluntary		

Contd...

♦ Yes	1	2
♦ No	0	0
• Audit of statements		
♦ By External auditor	1	2
♦ By Internal auditor	1	1
♦ No Audit	0	0
(c) Recognition of Social Costs		
• Determination of social costs	1	2
• Valuation of social costs	1	3
• Consideration in decision-making	1	1
3. Operational Aspects		
(a) Training to employees		
• Yes	1	2
• No	0	0
(b) Internal reward system		
• Financial	1	2
• Non-financial	1	1
• No system	0	0
(c) External recognition for environmental protection		
• Yes	1	2
• No	0	1
(d) Voluntary environmental protection		
• Yes	1	2
• No	0	0

Here, the companies were grouped into two categories: 'most polluting industries' and 'less polluting industries'. Secondly, on the basis of product manufactured, the sample companies were divided into 11 industries. Had both these classifications been used in one equation, there would have been a high degree of multicollinearity among independent variables. Hence, only one classification system was used in a particular equation at a time. Accordingly, two types of equations were formulated namely, Type I and Type II equations. In Type I equation, industry categorization according to the polluting nature of industry was done. In

Type II equation, industry classification according to nature of product manufactured was used. Both the equations were again estimated twice – firstly, by using equal weights in measuring EDS and EP and secondly, using unequal weights in the measurement of these two variables. All other variables in these equations had same meaning. These equations along with the analyses are given in the next Section.

6.5 Analysis of Regression Results

This Section presents an analysis of the estimated regression equations. The results have been given in two sub-sections. While sub-section 1 gives estimation of Type I regression equations, sub-section 2 presents testing of Type II regression equations.

6.5.1 Estimation of Type I Regression Equations: As mentioned before, in Type I equations, companies were grouped on the basis of polluting nature of their industries. The Type Ia equation is as follows:

Type Ia equation:

$EDS = \square + \square_1 D_S + \square_2 D_{MP} + \square_3 D_{FA} + \square_4 D_{BH} + \square_5 S + \square_6 P + \square_7 E + \square_8 D/E + \square_9 EP + e$ where

EDS=Environmental disclosure score (equal weights)

$\square$=constant

$\square_1$ to $\square_9$=regression coefficients

D_S = Dummy for sector

It is 1 if public sector company; 0 otherwise

D_{MP}=1 if most polluting industry; 0 otherwise

D_{FA}=1 if the company has foreign association; 0 otherwise

D_{BH}=1 if the company belongs to a large business house; 0 otherwise

S=Size

P=Profitability

E=Exports to sales ratio

D/E=Debt-equity ratio

EP=Environmental Performance (equal weights)

e=error term

In Type I_b equation, all the variables were same except two variables namely EDS and EP, which were measured by using unequal weights. The estimated Type I equations are given in Exhibit 6.11.

The Exhibit reveals that both the equations were significant at 1 percent level of significance. When equal weights were used to measure EDS and EP, all the independent variables taken together explained 40.1 percent variation in EDS. When adjusted for degrees of freedom, the variation explained (as indicated by adjusted R^2) reduced to 33.4 percent. In Type I_b equation (using differential weights), the explained variation as indicated by adjusted R^2 was 32.8 percent. The Exhibit further depicts that in both the equations, EDS of the public sector companies was significantly higher as compared to the private sector companies.

Exhibit 6.11: Estimated Type I Equations (Original)

Equation	Constant	D_S	D_{MP}	D_{FA}	D_{BH}	S	P	E	D/E	EP	R^2	R^2	F-value	Sig.level
Type I_a	–.0482	.110 (2.211)**	.0758 (2.193)**	.0491 (1.083)	.0461 (1.282)	.000003 (2.031)**	.0010 (1.737)*	.0006 (.558	–.014 (–1.41)	.0167 (3.33)***	.401	.334	6.018	.000
Type I_b	–.0544	.127 (2.49)**	.0887 (2.54)**	.0477 (1.04)	.0457 (1.25)	.000003 (1.98)*	.0011 (1.96)*	.0006 (.56)	–.013 (–1.21)	.0092 (2.74)***	.395	.328	5.878	.000

Note: 1. T-Values Are Given in the Parentheses 2. *Significant at 10 Percent Level of Significance **Significant at 5 Percent Level of Significance *** Significant at 1 Percent Level of Significance

The dummy variable used for public sector companies was positive and significant at 5 percent level in both the equations. Similarly, the EDS of companies in 'most polluting industries' was significantly higher than EDS of those in 'less polluting industries'. The coefficients of other two dummy variables (depicting foreign association and association with large business houses) were not statistically significant. Regarding quantitative variables in the equation, the variables 'size', 'profitability' and 'environmental performance' were found to be significant in explaining variation in environmental reporting practices of the sample companies. All the variables (except debt-equity ratio) had expected signs. Debt-equity ratio (though not significant) was found to be negatively related to the EDS, contrary to the expectation. It seems that in actual practice, companies having high debt-equity ratio, disclose lesser information on environment. Fear of action from regulatory authorities or loss of credit seem to be the reasons for this.

An attempt was made to test the presence of superfluous variables [28] in these equations. For this, each one of the variables was dropped one by one from the estimated regression equations and its effect on the related statistics was studied. The results [29] revealed the following:

- R^2 decreased in most of the cases. However, it increased after removal of one variable 'exports' and remained almost same after deletion of variable 'foreign association'. F-value also increased with the removal of these two variables. [30] The t-values of all the other variables did not change to the extent to change their significance.
- In all other cases, removal of the variable led to decrease in adjusted R^2. In some cases, the significance of regression coefficients of other variables also changed.
- Signs of the parameters did not change in all the cases.

On the basis of above analysis, two variables (foreign association and exports) appeared to be superfluous. They were also found to be insignificant in the analysis. The correlation matrices [31] showing correlation among independent variables for Type I_a and Type I_b equations also showed that 'foreign association' and 'exports' were significantly correlated to variable 'sector'. Thus, the deletion of these variables was also justified on the basis of criteria of 'multicollinearity among independent variables.' [32] Hence, these two insignificant variables were removed from the Type I_a and Type I_b equations. The Final Type I_a and Type I_b equations thus, derived are given in Exhibit 6.12.

Exhibit 6.12: Estimated Type I Equations (Final)

Equation	Constant	D_S	D_{MP}	D_{FA}	S	P	D/E	EP	R^2	R^2	F- value
Type I_a	–.0159	.0895 (1.919)*	.0759 (2.22)**	.0321 (.942)	.000003 (2.07)**	.00103 (1.83)*	–.0151 (–1.48)	.0157 (3.19)***	.389	.338	7.558***
Type I_b	–.0202	.106 (2.23)**	.0886 (2.57)**	.0319 (.922)	.000003 (2.01)**	.00117 (2.04)	–.0135 (–1.296)	.00852 (2.60)***	.384	.332	7.401***

Note: 1. T- Values Are Given in the Parentheses 2. * Significant At 10 Percent Level of Significance ** Significant At 5 Percent Level of Significance *** Significant At 1 Percent Level of Significance

An examination of Type I_a equation (Final) revealed that the regression equation was significant at 1 percent level of significance. All the independent variables explained 38 percent variation in the EDS percentages of selected Indian companies, which was reduced to 33.8 percent, when adjusted for number of degrees of freedom. When unequal weights were used to measure EDS and EP, almost same results were achieved. The variation explained (as indicated by adjusted R^2) was slightly less i.e. 33.2 percent here. The results provided sufficient evidence to support the hypothesis H_1, i.e., environmental information disclosure by the public sector companies is significantly higher as compared to the private sector companies. Environmental disclosures by 'most polluting' companies was also found to be significantly higher than that of 'less polluting' industries. This shows that environmentally sensitive industries have a tendency to disclose more information on their EP. This reporting strategy is generally used to reduce company's potential cost in terms of stringent environmental legislation, action from regulatory authorities or criticism from society. Kolk et al., in a study of 250 largest firms in the world observed that whereas environmental reporting in most of the less-polluting sectors was found to be just starting, there was an increase in number of externally verified environmental reports by companies operating in environmentally sensitive sectors. [33]

O'Dwyer in a study of 83 listed Irish companies also found predominance of environmental disclosures among companies in more environmentally sensitive sectors. They linked it to the desire of the companies to appear to be concerned about environmental issues. [34]

'Size', 'profitability' and 'EP' emerged as important explanatory variables. The results indicate that big and more profitable companies provide more information on environment than small and less profitable companies respectively. Reporting certain information voluntarily involves cost that cannot be afforded by small companies. A study by Gray et al. (2001) found that larger and more profitable companies tend to have a greater amount of each type of social and environmental disclosure. [35] Another study by Mahmud and Others (1994) suggested that big and listed companies have a tendency to provide more information than required by standards and legislation as to maintain the demand of their shares. Non-disclosure may be interpreted by investors as 'bad news and this could adversely affect the firm value. [36]

The results also showed that companies having better EP

reported more on environment than companies which did not perform well. It is obvious that a company which takes various steps to prevent and control pollution or to conserve environment (whether statutorily required or voluntarily taken) and takes care of environment in making crucial business decisions, tries to report this information to the public. The company tries to enhance its goodwill by displaying that it is a good corporate citizen. The results, therefore, provide evidence in supporting the hypothesis that companies having better EP disclose more on environment than poor environmental performers. As compared to this, some studies (e.g., Ingram and Frazier, 1980; Jaffar et al., 2002) found that poor environmental performers provide detailed information at strategic or important locations in the annual reports. [37]

Thus, 'sector', 'industry', 'environmental performance', 'size', and 'profitability' were found to be significant variables in explaining environmental reporting by the sample companies.

6.5.2 Estimation of Type II regression equations: In Type II regression equations, industry classification on the basis of 'product manufactured' was taken as one of the independent variables instead of 'most polluting' criteria for classifying industries used in Type I equations. The sample companies were divided into 11 industries (broadly using BSE classification). For incorporating 11 industries in the equation, 10 dummy variables were used. The omitted industry was 'other industries'. The Type IIa regression equation used here was as follows:

Type IIa equation:

$$EDS = \square + \square_1 D_S + \square_2 DI_1 + \square_3 DI_2 + \square_4 DI_3 + \square_5 DI_4 + \square_6 DI_5 + \square_7 DI_6 + \square_8 DI_7 + \square_9 DI_8 + \square_{10} DI_9 + \square_{11} DI_{10} + \square_{12} D_{FA} + \square_{13} D_{BH} + \square_{14} S + \square_{15} P + \square_{16} E + \square_{17} D/E + \square_{18} EP + e$$

Where, DI_1-DI_{10} were 10 dummy variables used to incorporate 11 industries in the equation. All other variables had same meaning as given in Type Ia equation. Here again equal weights were used to measure EDS and EP. In Type II_b

equation, unequal weights were used to measure these variables. The estimated Type II_a and Type II_b equations, using equal and unequal weights for measuring EDS and EP respectively, are given in Exhibit 6.13.

The Exhibit reveals that inclusion of this kind of industry classification did not improve the results. Though the variation explained by all the variables taken together, as indicated by R^2 increased to 47.2 percent and 45.6 percent in Type II_a and Type II_b equations respectively; it decreased significantly to 34 percent and 32 percent when adjusted for degrees of freedom.

Exhibit 6.13: Estimated Type II_a and II_b equations

Equation	Constant	D_S	DI_1	DI_2	DI_3
II_a	–0.0289	.0699 (1.173)	–0.045 (–.419)	.0326 (.423)	.0916 (.748)
II_b	–.0422	.0970 (1.57)	–.051 (–.464)	.0411 (.519)	.0953 (.770)

Contd...

DI_4	DI_5	DI_6	DI_7	DI_8	DI_9	DI_{10}
–.101 (–1.469)	.0757 (1.265)	–.02 (–.351)	.0205 (.355)	.0671 (.774)	.147 (2.1)	.168 (1.37)
–.111 (–1.57)	.0825 (1.34)	–.0246 (–.419)	.0377 (.627)	.0558 (.629)	.134 (1.86)	.170 (1.33)

Contd...

D_{FA}	D_{BH}	S	P	E	D/E
.050 (1.10)	.0358 (.946)	.000002 (.971)	.00041 (.584)	.00104 (.899)	–.0146 (–1.121)
.0489 (1.05)	.0379 (.976)	.000002 (.920)	.00053 (.740)	.00103 (.863)	–.0127 (–.943)

Contd...

EP	R^2	R^2	F-value	Sig. level
.0173 (3.20)***	.472	.34	3.579	.000
.0101 (2.74)***	.456	.320	3.348	.000

Note: 1. T-Values Are Given In The Parentheses 2. * Significant At 10 Percent Level of Significance ** Significant At 5 Percent Level of Significance *** Significant At 1 Percent Level of Significance

EDS of only I_9 (i.e. oil, gas and petrochemicals industry) was found to be significantly higher than EDS of 'omitted industry'.

No other variable was found to be significant (except environmental performance) in explaining variation in environmental disclosure practices of the sample companies; though overall, both the regression equations were significant at 1 percent level. There might be two reasons for this–firstly, high degree of correlation among these industries and secondly, small number of companies in some of the industries (e.g., in I_1-coal, mining and cement industry only three companies were there). An attempt was therefore, made to improve the results by grouping the industries on the basis of some similarity in nature. All the companies were classified in five industrial groups in the final analysis, which were:

- I_a=Iron, steel and other metals
- I_b=General engineering, automobiles, electrical products
- I_c=Chemical, Fertilizers and Pharmaceuticals
- I_d=Oil, Gas, Petrochemicals and Power Generation (mainly Infrastructure related industries)
- I_e=Others

I_b was taken as 'omitted industry' as it is considered 'less polluting' as compared to other industries. All other variables (except industry groups) had same meaning as used in equations IIa. The estimated equations, namely Type II_c and Type II_d, using grouped industry classification are given in Exhibit 6.14.

An examination of Type II_c equation revealed that the regression equation was significant at 1 percent level of significance. All the independent variables explained 36.5 percent variation in EDS (as indicated by adjusted R^2). EP was the most important independent variable in explaining reporting practices on environment. EDS in 'Iron, steel and other metals' was significantly higher as compared to omitted industry (i.e., 'general engineering, automobiles and electrical products'). Similarly EDS in 'oil, gas, petrochemicals' and 'power generation' was significantly higher as compared to EDS in 'omitted industry'.

Exhibit 6.14: Estimated Type IIc and Type II d Equations (Original)

Equation No.	Constant	D_S	D_{FA}	D_{BH}	S
II_c	.0856	.0844 (1.66)*	.0543 (1.23)	.0316 (.871)	.000002 (1.24)
II_d	–.105	.112 (2.14)**	.0516 (1.13)	.0342 (.915)	.000002 (1.27)

Contd...

P	E	D/E	EP	DI_a	DI_c
.00078 (1.36)	.0008 (.718)	–.0115 (–1.15)	.0182 (3.77)***	.117 (2.35)**	.0621 (1.31)
.00094 (1.58)	.0008 (.690)	–.0093 (–.901)	.0106 (3.17)***	.13 (2.54)**	.086 (1.75)*

Contd...

DI_d	DI_e	R^2	R^2	F-value
.185 (3.26)***	.068 (1.68)	.45	.365	5.31***
.180 (3.07)***	.075 (1.75)*	.431	.344	4.92***

Note: 1. T-Values Are Given In The Parentheses 2. * Significant At 10 Percent Level Of Significance ** Significant At 5 Percent Level Of Significance *** Significant At 1 Percent Level of Significance

There seems to be two reasons for this: firstly, a large number of companies in these industries belonged to public sector, where environmental reporting was found to be more as compared to private sector and secondly, these were 'environmentally sensitive' industries. Stringent government regulations and pressures from governmental and non-governmental organizations compel these companies, not only to take steps to protect environment, but also to demonstrate their environmental performance to the public at large.

EDS in the 'public sector' companies was again found to

be significantly higher as compared to EDS in the 'private sector' companies. An examination of Type II_d equation revealed almost same results. The equation was significant at 1 percent level of significance. 'Sector' and 'EP' emerged as significant explanatory variables. 'Beta' coefficients of dummy variables used for all four industrial groups were positive and significant. This shows that in all these four groups, EDS was significantly higher as compared to EDS in 'omitted industry'.

An attempt was again made to test these equations for the presence of superfluous variables. For this purpose, each one of the independent variables was dropped one by one from the estimated regression equations. The equations so obtained and the related statistics are tabulated and shown in Appendix V. The results revealed that adjusted R^2 decreased in all the cases except when variables 'exports' and 'business house' were dropped. F-ratio decreased in some of the cases, but it increased when these two variables were dropped. 't-value' of the variables changed in some of the cases significantly. [38] Signs of parameters did not change. A study of second equation revealed quite similar results. Here deletion of one more variable i.e. debt-equity ratio increased R^2. But this variable could not be said to be superfluous as its deletion changed the 'beta coefficients' of some other variables to the extent that they became statistically insignificant. Therefore, only 2 variables (business house and exports) appeared to be superfluous. They were also insignificant in the equation and hence, they were dropped from the estimated equations. The regression equations, finally obtained are given in Exhibit 6.15.

A close scrutiny of these equations revealed that both the equations were significant at 1 percent level of significance. All the independent variables explained 42 to 44 percent variation in the dependent variable (as shown by R^2), which was reduced to 35-37 percent when adjusted for degrees of freedom. An examination of Type II_c (final equation) further revealed that 'environmental performance' was the most important variable in explaining variation in EDS. Moreover, EDS in 'Iron steel and other metals' industry and 'oil, gas, petrochemicals and power generation' was significantly higher as compared to EDS in 'omitted industry'.

Exhibit 6.15: Estimated Type IIc and Type IId Equations (Final)

Equation	Constant	D_S	D_{FA}	S
II_a	.0631	.0602 (1.36)	.0466 (1.098)	.000002 (1.20)
II_b	–.0796	.0861 (1.89)*	.0429 (.984)	.000002 (1.23)

Contd...

Equation	P	D/E	EP	DI_a	DI_c
II_a	.00078 (1.36)	–.0103 (–1.045)	.0181 (3.77)***	.126 (2.59)**	.0642 (1.37)
II_b	.00093 (1.58)	–.00789 (–.788)	.0104 (3.16)***	.139 (2.79)***	.0877 (1.81)***

Contd...

Equation	DI_d	DI_e	R^2	R^2	F value
II_a	.193 (3.47)***	.653 (1.59)	.442	.372	6.328***
II_b	.188 (3.28)***	.0713 (1.69)*	.423	.350	5.856***

Note: 1. T-Values Are Given In The Parentheses 2. * Significant At 10 Percent Level Of Significance ** Significant At 5 Percent Level Of Significance *** Significant At 1 Percent Level Of Significance

When unequal weights were used in measuring some variables, slightly better results were obtained. An examination of Type II_d (final equation) revealed that EDS in 'public sector' companies was significantly higher as compared to EDS in 'private sector' companies. Environmental performance again emerged as the most important quantitative variable in explaining variation in EDS. EDS in all the 4 industrial groups included in the equation were significantly higher as compared to EDS in 'omitted industry'. A study by

Davut Ince also found that industry and size effect appear to be existed in stakeholder related social and environmental information disclosure in Environmental Policy Statements. [39] Similarly, Gray et al. (2001) found that measure of industry classification had significant influence on the results showing impact of various corporate characteristics on environmental disclosure. However, this study recommended for a more systematic analysis of the various factors under different specifications of industry classification which may well prove illuminating. [40]

6.6 Summary

This Chapter examined environmental disclosure practices in the annual reports of the selected large manufacturing companies in India. The analysis indicated that there were broad variations in the format and content of corporate environmental disclosures. Though all the companies gave in detail statutorily required information on energy conservation, voluntary reporting was not the order of the day. Some companies reported on environmental policies, legal compliance, pollution control etc., but it was not a regular feature. Only a negligible number reported information on environmental costs, budgets, liabilities or social costs. Majority of the sample companies gave only positive information in the annual reports, like steps taken for environmental protection, pollution control, new innovations in the area and awards obtained. However, a few companies disclosed negative information on environment like polluting nature of the processes used and accidents in a year. Usually, the environmental information was found in the Director's Report, rather than in the financial statements or footnotes. The companies generally disclosed only qualitative/descriptive information on the environment in the annual report. Only a few companies reported quantitative/financial figures on the issue and the information provided was generally brief and lacked specific details.

Thus, majority of the companies reported only qualitative, statutorily required and positive information on selected environmental issues. It was felt that in the absence of any accounting standard or guidelines on the issue, both quantity and quality of disclosure was quite limited. Moreover, there was no consistency in these reporting practices over the years. The findings of this Chapter substantiate the hypothesis that companies in India do not give elaborate reporting on environmental issues.

An attempt was also made to examine the impact of selected company and industry-related variables on environmental disclosure practices of the sample companies. Multiple regression analysis was used for this purpose. Two types of regression equations were developed. These were Type-I and Type-II equations using different criteria for classifying industry. In both the cases, further analysis was done using equal and unequal weights to measure some of the variables.

In Type I regression equations using 'polluting nature of industry' criterion for classifying the companies, 'size', 'profitability' and 'environmental performance' were found to be significant variables in influencing the extent of environmental reporting. Environmental disclosure by the 'public sector' companies was found to be significantly higher as compared to 'private sector' companies. Further, EDS of companies in 'most polluting industries' was found to be significantly more as compared to EDS of companies in 'less polluting industries'.

Type II regression equations were used to empirically test differences in environmental reporting practices of the industries classified on the basis of 'nature of product manufactured'. When this classification was used, it was found that though overall regression equations were significant, the results failed to capture industry effect on disclosure practices of the companies in different industries (may be due to small number of companies in some of the industries). Then, all the companies were grouped into 5 industry groups and better results were obtained. Environmental performance again emerged as the most important explanatory variable. When equal weights were used, EDS in only 2 industrial groups was found to be significantly higher as compared to 'omitted industry' group. However, when unequal weights were used, EDS in the 'public sector' companies and all the 4 industry groups included in the equation were found to be significantly higher than their respective omitted categories. Thus, in this case measurement with unequal weights was found to be better as against measurement with equal weights, in explaining difference in environmental reporting practices in different industries.

To conclude, environmental reporting in India is still at infancy stage. Formal corporate environmental reporting is not directly regulated in India and because of this reason, companies in India enjoy significant freedom to choose which issues and aspects of environment to be disclosed to the public. The level of environmental disclosure is highly explained by sector and nature of industry to which a company belongs. Moreover, it is significantly affected by environmental performance of a company. Companies that have better environmental performance are expected to provide more environmental information in the annual reports as compared to poor performers. The study also indicates existence of statistically significant 'size' and 'profitability' effect.

Notes

1. Some of these sources are mentioned as follows: UN-ICAI Seminar Report, *'Environmental Accounting and Reporting'*, conducted at New Delhi, (Aug. 2-4, 1999), pp. 6.9-6.11; (b) Environmental Disclosure Requirements of ICAEW, U.K., (1992); (c) CTC ISAR, 9th Session, Recommendations for Environmental Financial Reporting, in Gray R., *Accounting for the Environment*, Paul Chapman, U.K. (1998); (d) Deegan, C., "The Environment: An Accountants Dilemma", *Charter*, (August 1994), p. 50; (e) Gray R. and Bebbington J., *Accounting for the Environment*, Sage Publications, London, 2nd ed., (2001),

p. 248.

2. The finding that all the items of information were regarded as of having some importance by the CAs was not surprising. In fact, one of the criteria for including an item in the 'Index' was that it should be at least of 'some importance' to the stakeholders of the annual report. Including items with no expected importance would only lengthen the index and perhaps even waste the respondents' precious time.
3. As mentioned in the previous Chapter, the companies for the study were selected from Delhi Stock Exchange's list of top 500 companies in India. The primary information was supplied by 101 companies. Out of these 101 companies, annual reports of only 91 companies were examined because for others these reports were not available for all the three years.
4. Most of the existing studies have considered corporate annual reports for analysing environmental reporting practices (e.g., Borghini, 1998; Guthrie and Parker, 1983; Thomas, 2001; Gray et al., 2001; Jones, 2000). For detailed references, please see Chapter IV.
5. Tilt in a study conducted in 1994 found that annual report was considered by the companies to be the most appropriate medium for disclosing information on environment, demonstrated by their preference for it over other mediums. See Tilt A., "The Content and Disclosure of Australian Corporate Environmental Policies", Source: www.ssh.flinders.edu.au/commerce/researchpapers/aa.4.htm, last accessed on 30.1.2005.
6. The mean score of each item of information in the 'Index of environmental disclosure', as calculated in Exhibit 6.1, was used as unequal (differential) weight.
7. Tilt, in a study of content and disclosures of Australian corporate environmental policies, also found little evidence of any quantified targets or specific objectives in Australian annual reports. See Tilt (1994), *op. cit.*
8. As mentioned above, the companies are required to disclose information on conservation of energy in the annexure to the Director's Report. It has been observed that since companies have to report on this issue, they take it as an indirect legal compulsion to do something to conserve energy. It has been rightly said, "What is to be reported, is gets done." Reporting of

this item in the annexure has been regarded as disclosure of statutorily required measures. Information given elsewhere in the annual report has been considered as reporting of initiatives taken in this regard.

9. See Frost G.R., "An Investigation of the Introduction of Mandatory Environmental Reporting in Australia", 3rd Asia Pacific Interdisciplinary Research in Accounting (APIRA) Conference, (15-17 July, 2001), Adelaide.
10. See Larrinaga C., "An Evaluation of Mandatory Environmental Accounts in Spain", Source: visar.csustan.edu/papers/ larrinage 91.pdf, Last accessed on 30.1.05.
11. See Jones K., *Study of Environmental Reporting by Companies*, Commission of European Communities, DG Environment, Centre for Environmental Informatics, University of Sunderland, (March 2000).
12. See Moneva J.M. and Llena F., "Environmental Disclosures in the Annual Reports of Large Companies in Spain", *The European Accounting Review*, 9:1 (2000), pp. 7-29.
13. See O'Dwyer B., "The Ponderous Evolution of Corporate Environmental Reporting in Ireland: Recent Evidence from Publicly Listed Companies", *Corporate Social Responsibility and Environmental Management*, 10 (2003), pp. 91-100.
14. See Roy, M.N., 'Environmental Disclosure in the Corporate Annual Reports', in Rao P.S.B. and Rao P.M., *Environment Management and Audit*, Deep and Deep Publications Pvt. Ltd., Delhi, (2000), pp. 228-244.
15. See Jaffar R., Iskandar T.M. and Muhamad M., "An Investigation of Environmental Disclosure in Malaysia", Source: www.aaanz.org/web2002/ accepted0.020-papers/jafar.pdf, as last accessed on 30-1-05.
16. See Jones K., *op. cit.*
17. For measuring length of environmental disclosures 'page' was used as a unit of measurement. It was noted that annual reports of different companies differed in page size and font used. Thus, they were not comparable. Another method which could have been used for measuring the length of disclosure was word or line count. But this could not cover disclosure through charts or tables which was quite important. Therefore, number of pages covered was used as a measure of length. The results should be viewed keeping in mind limitations of this measure.

Some existing studies (e.g., Jaffar et al., 2002; Eresi, 1996; Roy, 2000 and Stratos Institute, 2000) also measured quantity of reporting on the basis of number of pages used to report environmental issues. However, Deegan and Gordon, 1996; and Deegan and Rankin, 1996 used word count to measure the quantity of positive and negative disclosures made by Australian corporations. Tilt, 1999 used number of sentences as a measure of disclosure. For detailed references of these studies, see Chapter IV, Section 4.2.

18. See PIRC (1996), adopted from, Pramanik A.K., '*Environmental Accounting and Reporting*', Deep and Deep Publications, (2003), p. 351.
19. See Roy, *op. cit.*
20. Gujrati D. N., *Basic Econometrics*, McGraw Hill Book Company, New York, (2003), p. 18.
21. "Even if the relationship is not linear, when the relevant range of operation is small, the linear form may adequately represent the true functional form". See Rao P. and Miller R.L., *Applied Econometrics*, Prentice Hall of India Pvt. Ltd., New Delhi, (1972), p. 2.
22. For details, see Kmenta J., *Elements of Econometrics*, Macmillan Publishing Company, (1971), p. 348.
23. Gujrati D.N., o*p. cit.*, p. 298.
24. *Ibid.*, p. 323.
25. 6 KPMG, *'International Survey of Corporate Sustainability Reporting'*, (2002), Source: http:// www.wimm.nl/publicates/KPMG2002.pdf.
26. As observed in Chapter V, a large majority of the CAs and the responding executives felt that environmental information is very important in internal decision-making. 57 percent of the executives informed that their companies prepare internal environmental accounts.
27. a) In a study by Jaffar, Iskandar and Muhamad (2002), EP was measured on the basis of legal action taken on the companies by the Department of Environment (DOE) for non-compliance of environmental regulations. More specifically, the EP was measured based on severity of environmental problems caused by a company. See Jaffar R., Iskandar T.M. and Muhamad M., "An Investigation of Environmental Disclosure in Malaysia", Source: www.aaanz.org/web2002/ accepted0.020-

papers/jafar.pdf, as last accessed on 30-1-05. Some other studies (e.g., Deegan and Rankin, 1996) measured companies' non-compliance of environmental regulations based on the courts action taken on companies that caused problems to the environment. See Deegan C. and Rankin M., "Do Australian Companies Report Environmental News Objectivity?", *Accounting, Auditing & Accountability Journal*, Vol. 9, No. 2 (1996), pp. 50-67. The present study had not used these measures as they were mainly negative in nature. A more positive, primary data based approach was used to measure EP in the present study.

28. A superfluous variable is one whose exclusion or inclusion in regression equation does not alter the other coefficients' estimates or their standard errors 'significantly', that is, when such variable is added or deleted in a regression equation, the 'damage' to the regression coefficients of the other variables is not operationally significant. The presence of superfluous variable is judged on the basis of the following guidelines: Its deletion from the regression equation does not decrease R^2 significantly to decrease R^2. The exclusion of a superfluous variable does not alter the estimates of other coefficients or their standard errors significantly. However, if discarding of a variable changes the regression coefficients of other variables, then it cannot be superfluous, even if its deletion increases R^2. See Rao P. and Miller R.L., *op. cit.*, pp. 35-40.
29. Refer to Appendix V for Exhibits relating to testing for superfluous variables and Multicollinearity.
30. An equation having higher value of F-ratio is regarded better than one having its lower value.
31. Refer to Appendix V.
32. Though correlation between some other variables was also found to be significant, they were not deleted from the equation, keeping in mind their significance in explaining variation in EDS and theoretical justification. Hence, the final results must be viewed considering presence of multicollinearity in some variables.
33. Kolk A., Walhain S. and Wateringen S.V., "Environmental Reporting by the Fortune Global 250: Exploring the Influence of Nationality and Sector", *Business Strategy and the Environment*, 10, (2001), pp. 15-28.

34. O'Dwyer, *op. cit.*
35. See Gray R., Jarad M., Power D., and Sinclair C.D., 'Social and Environmental Disclosure and Corporate Characteristics: A Research Note and Extension', *Journal of Business Finance and Accounting*, 28(3) & (4), (April/May 2001), pp. 327-356.
36. Mahmud H., Tan L.M., and Adams M., "Voluntary Disclosure in an Emerging Capital Market: Some Empirical Evidence from Companies Listed on the Kuala Lumpur Stock Exchange", *The International Journal of Accounting*, 29, (1994), pp. 334-351.
37. See a) Jaffar et al., *op.cit*; and b) Ingram R. and Frazier K., "Environmental Performances and Corporate Disclosure", *Journal of Accounting Research*, Autumn, (1980), pp. 614-622. The contrary results might have been obtained due to difference in methods used to measure EP. As mentioned before, Jaffar et al. measured EP on the basis of legal action taken on companies by Department of Environment (DOE) for non-compliance of environmental regulations. A higher score indicated that the company was more environmentally problematic. They found that companies that performed badly in terms of EP, provided detail information on the issue to reduce potential political cost. In the present study, a higher score indicates better EP in terms of contribution towards protection of environment and the results show that better environmental performers disclose more environmental information than poor performers. It might also be possible that a company which was environmentally problematic in the past has now taken various steps to reduce its environmental problems and improve its environmental performance.
38. For example, when variable D_{Id} was removed, the 3 variables sector, size and profitability became significant. This might be due to high degree of multicollinearity present in these variables. The variable was retained in the equation as its deletion decreases R^2 significantly.
39. See Ince D., "Determinants of Social and Environmental Disclosures in the UK Companies in Environmental Policy Statements", Source: http://les.man.ac.uk/IPA97/papers/ince119.pdf.last accessed on 12-3-04.
40. See Gray et al., *op. cit.*. The present work attempts to incorporate their suggestions in the analysis.

7

Summary, Conclusions and Policy Recommendations

Accounting and reporting for the environment has become an important dimension of corporate external information system in 21st century. The increasing pressure on companies to be responsible to the society has influenced them to operate in an environmentally responsible manner. 'The impact of environmental performance of an enterprise on its financial health' has become a matter of growing concern to investors, creditors, government and public at large. As various stakeholders demand greater disclosure of environmental impacts and performance, a large number of companies, all over the world, have started reporting on these issues. In many countries, disclosure of some environmental information (in or outside the financial statements) has also been made mandatory. However, various research findings have shown that in the absence of any accounting standard or guideline on the issue, this disclosure generally lacks quality and consistency in reporting. A more disciplined approach, therefore, is required for accounting and reporting of environmental information to improve its quality.

This book aimed at presenting various theoretical and legal considerations relating to environmental accounting and reporting (EAR). An analysis of EAR practices of Indian corporate sector was made in the book. An attempt was also made to examine existence of corporate EAR expectation gap in India. This Chapter presents important findings of the study and makes some recommendations. It is hoped that these observations and suggestions would be of immense use to investors, creditors, employees, government and other stakeholders. They would also help the companies and management in handling 'green matters' effectively and efficiently in the financial statements.

7.1 Observations

In recent years, environmental accounting has become a strong branch of accounting. The term environmental accounting can be explained at different levels. National level environmental accounting is a method of keeping accounts of natural resources and other environmental elements. It involves valuation of natural resources, measuring income therefrom, keeping records of costs relating to them, estimating them and providing depreciations on them. On the other hand, at corporate level, environmental accounting refers to the measurement and communication of information on the environmental responsibility performance of an entity to the interested parties. It is a part of corporate social responsibility and attempts to evaluate impact of organisational activities on environmental resources. While environmental financial accounting deals with collection, measurement and reporting of environmental information for external reporting purposes; environmental management accounting is concerned with presenting environmental data to management for internal decision-making. The book examined the concept of environmental accounting at the corporate level only.

It was observed that accounting of environmental information and its subsequent disclosure to various stakeholders is required to help them in making various relevant decisions. Information generated by environmental accounting system assists management in better decision-making. It helps in gaining competitive advantage by minimising adverse environmental impacts through improved designs, products or processes. Also, by motivating to reduce the use of and the expenditure on natural resources and the cost and impacts of disposing of waste- the application of environmental accounting might help increase the profitability and productivity of organizations. By considering social and environmental externalities, an organization can start to assess its long-term sustainability. Reporting of this information to various interested stakeholders also promotes improved performance by forcing to reduce adverse environmental impacts of the company for the purpose of communicating them to various stakeholders. Thus, accountants and managers need to be aware of environmental issues. They should also consider influence of these issues upon both internal and external reporting.

Over the past few decades, a significant body of accounting and finance literature has witnessed an increase in the demand for company's environmental information. Investors, creditors, government and other stakeholders now demand disclosure of information regarding emissions (even if legal), failure to comply with environmental regulations and potential liability to environmental remediation requirements. It helps these stakeholders in assessing environmental risks and environmental performances of the companies and assists them in making informed decisions. They consider such information relevant to their decisions—and thus financially material. However, the availability of information on environmental issues has not kept pace with its growing financial significance.

The analysis of accounting standards and guidelines in different countries showed that the accounting standards setters at the international and national levels have generally chosen to avoid dealing directly with the topic of environmental issues in financial reporting. Till date, only a few international standards exist concerning how environmental issues may be treated in annual reports. The recommendations of the UN ISAR group and the EU represent worthy attempts by non-accounting organizations to fill the gap and should be taken forward by various accounting standard setters.

Till date, in India, there is no financial reporting or auditing standard dealing explicitly with social, environmental or sustainability related issues. The current financial accounting framework in India appears to be ill-equipped to provide the information required by various internal and external stakeholders on environmental costs and liabilities, environmental risks involved in their investments, social costs or externalities and steps taken by companies to reduce global warming and improve sustainability. This shows that the accounting profession in India has yet to display significant role in sustainability. It should contribute in this area by developing an environmental and sustainability reporting framework on the lines of those developed by FEE and GRI. Environmental reporting should be made mandatory in India. Moreover, in order to provide relevant, reliable and comparable information to the interested stakeholders, it is necessary that only duly audited information is provided to them. This gives rise to the need for independent environmental information audit.

Disclosure of social and environmental information has been an area of substantial research in the past few decades. Since 1990, a number of studies have been made to examine the environmental reporting practices of the companies. A review of existing studies showed that though some international surveys (e.g., by KPMG and SustainAbility at different points of time) examined environmental information disclosures by concerns from different countries, most of the country-specific works focussed on the green reporting practices of corporations mainly from Europe and USA. Empirical evidences from the developing countries were very small in number. Most of the existing studies analysed communication of environmental information by companies in general, without concentrating on quality of reporting. Very few studies had examined the impact of various corporate and industry-related variables on ecological disclosures. A preponderance of existing works were based on secondary data and lacked insight into various conceptual and theoretical considerations involved in accounting and reporting on environment.

Environmental accounting and auditing areas are yet to be explored by researchers (particularly in India). Most of the existing attempts on environmental accounting were quite general in nature and did not focus on the specific issues involved in accounting for environmental information (like expensing and capitalisation of environmental costs, depreciation of environmental assets, and recognition and measurement of environmental liabilities). Only a few surveys (e.g., KPMG, 2002; ACCA, 2002) addressed the issue of verification of environmental information reported by the companies. There seems to be no reported research work covering all the three aspects, namely environmental accounting, reporting, and auditing. In fact, there is a need to cover these issues together as these are interrelated and form an important part of a company's environmental management system. The role of accounting in improving environmental performance of an organisation is also required to be understood. The present work has attempted to bridge these research gaps.

7.2 Corporate EAR Expectation Gap in India

Review of existing literature showed that environmental reporting in India is typically deficient and not of a standard to satisfy the information needs of various classes of report users. The present study tried to explore whether a potential demand/supply imbalance is due to differing perceptions between report users and preparers as to the various matters associated with EAR and importance of environmental information to the users' decision-making processes. For this purpose, the study surveyed the attitudes of senior executives in 101 large manufacturing companies in India (the preparers group) on various issues relating to EAR. An attempt was also made to obtain information on EAR practices of these companies from the executives. In addition, opinions of the selected individual Chartered Accountants (CAs), who were annual report users in different capacities, were obtained regarding their expectations about corporate EAR practices in India. An attempt was made to examine whether an EAR expectation gap exists within India.

The required information was obtained by using two structured questionnaires. The first questionnaire was used to obtain opinions of the CAs (selected from the directory prepared by ICAI) on various issues relating to EAR. The second questionnaire was used to examine EAR practices followed by the companies in India. The companies were selected from the Delhi Stock Exchange's List of top 500 companies in India. Manufacturing companies from both public and private sectors were approached for the survey. 101 companies supplied the desired information, which were used in the analysis. The findings of the primary data analysis have been presented in this part;-

7.3 Environment and Environmental Accounting

Generally, the CAs felt that 'environment' is a big challenge for business these days and companies must establish an efficient system of environmental management to take care of it. Establishment of a good EMS helps in maintaining clean and green environment around the factories, controlling pollution emission and thereby meeting legal standards in this area. The executives also found environment as one of the most important challenges faced by them. An examination of actual practices showed that about 90 percent of the companies already had EMS in their organizations, while some companies were planning to introduce it in near future.

The CAs as well as the corporate executives were of the view that the existing financial accounting framework is not comprehensive enough to deal with specific environmental problems. They realized the need for a separate standard on the issue to effectively deal with specific matters involved in accounting for the environment. The qualified accountants were of the view that a large number of companies do not prepare environmental accounts because these are not mandatory. 'Lack of accounting standards' and 'ignorance of benefits of environmental accounting' were other reasons considered important by some of them. They felt that companies should prepare environmental accounts mainly because it would help the companies in complying with various legislations. Moreover, adoption of environmental accounting system will facilitate decision-making by management and provide better estimates of the cost of production.

An analysis of environmental accounting practices of the sample companies also revealed that a large majority of the companies did not prepare environmental accounts for external reporting mainly because 'it is not mandatory'. 'No specific standards on the issue' was cited as another reason for non-preparation of these accounts. Thus, a significant expectation gap exists in this area. While users want that companies should prepare environmental accounts, companies do not prepare these accounts mainly because this is not required by law. However, 57 percent of these companies informed that they generated some environmental cost information for internal decision-making. The study found popularity of the concept of environmental accounting in India only at the management accounting level.

A large majority of the respondents considered environmental cost information very important as it provides a significant input in making strategic business decisions like in environmental impact assessment, designing of various processes using eco-friendly technologies, and evaluating environmental performance of a company. However, some of them did not find usefulness of this information in 'planning cost reductions'. The respondents favoured segregation of combined business expenses into environmental and other business costs. They felt that this segregation would help in finding out the full environmental costs incurred by a business and it would facilitate decision-making. Most of them favoured development of some accounting standard or guideline for bifurcating these combined costs. However, some of them recommended use of industry guidelines or internal mechanism for this purpose in the absence of a separate standard.

By and large, the qualified accountants as well as the senior accounts executives preferred use of conventional financial accounting principles in treating environmental costs in the financial statements. Accordingly, environmental costs which do not give any future economic benefit should be charged to Profit and Loss Account in the same year in which these costs are incurred. If an environmental cost gives economic benefits to the company in future, it should be capitalised. However, about 60 percent of them were in favour of capitalisation of even those environmental costs that do not give any direct future economic benefit to the company, but are necessary for the survival or continuation of business or are incurred for social benefits. The capitalised environmental assets should be depreciated over a reasonable period of time on the basis of relevant accounting standards or provisions of company law. Most of the respondents felt that environmental assets should be tested for impairment from time to time due to rapidly changing laws and technologies. Wherever appropriate, these assets should be written down to their recoverable amounts.

The respondents preferred following of the existing accounting standard to treat environmental costs relating to a prior period. Accordingly, an environmental cost related to a prior period should not be treated as a prior period adjustment unless there is some fundamental accounting error. About three-fourths of the CAs felt that pollution permits and emission rights must be treated as intangible assets, provided these are given for a longer period. However, 71 percent of the executives were not in favour of this, as at present the expenditure incurred in acquiring these rights is a recurring expenditure for factories.

The CAs, in general, did not favour separate recognition of environmental benefits in the financial statements. They were of the view that most of these benefits relate to future and keeping in mind 'principle of conservatism', should not be recognised in the current year's accounts. 'Intangible nature of most of the environmental benefits', 'difficulty in measurement' and 'automatic recording of realised environmental benefits' were the other reasons cited by some of them. However, more than one-half of the executives favoured separate recognition of these benefits in the financial statements as it will 'highlight an organisation's commitment towards environment' and will also 'help in doing cost-benefit analysis'. Some of them felt that it will have motivational impact on management. These executives recommended separate recognition of current realised environmental benefits in the annual report or environmental statement. In addition, future/expected benefits can be mentioned in the footnotes/annexure to the financial statements.

The respondents generally supported recognition of environmental liabilities in the financial statements when there is an obligation (legal as well as constructive) on the part of a company to incur an environmental cost. When there is difficulty in estimating an environmental liability, a reasonable estimate of it should be provided. In rare circumstances, when it is not possible for a company to estimate environmental liabilities, this fact along with the reasons therefore, should be disclosed. For measurement of long-term environmental liabilities, most of the respondents preferred use of 'discounted present value of expected future costs using risk-free rate'.

A large majority of the CAs were of the opinion that companies should try to determine external adverse impact of their activities (social costs) and consider them in making strategic business decisions. Most of them also supported an attempt by companies to value these social costs. Regarding practices of the companies, while about four-fifths of them determined external adverse impacts of their activities; only 42 percent claimed that they also tried to value these external costs. Most of the companies considered these costs in making crucial business decisions (e.g., capital budgeting or corporate restructuring decision). Almost all the CAs and the senior managers were of the view that due to growing social and legal pressures, these external/social costs will become internal in future. Therefore, companies should adopt a proactive approach to control these social costs by protecting and conserving the environment; otherwise, they may have to bear penalties.

7.4 Environmental Reporting

The respondents generally felt that companies in India do not give environmental information because the disclosure may jeopardize confidentiality in the sensitive areas and thereby, may adversely affect their competitive position. A large majority of the CAs were in favour of mandatory disclosure of environmental information. About three-fourths of the executives were also of the same view. They felt that compulsory environmental information disclosures will increase seriousness in the issue and will lead to overall increase in environmental reporting by Indian companies.

The CAs recommended that a company should use environmental information for both internal decision-making and external reporting. They felt that companies must disclose qualitative as well as quantitative information on these issues. Most of them preferred disclosure of this information in separate environmental statements to 'highlight its importance' and to 'avoid any confusion'. Furthermore, they also approved disclosure of qualitative information in the Director's Report.

An examination of the actual practices revealed that the companies generally used environmental information for internal decision-making as well as for external disclosure. About one-half of the companies disclosed only qualitative information on environment in the annual reports. Only 38 percent of them reported both qualitative and quantitative information in these reports. The companies mainly used the Director's Report to supply environmental information to external stakeholders. Some companies also prepared separate environmental statements to communicate detailed information on environment. Some of the executives were of the view that disclosing these details in the main body of financial statements would serve a better purpose. However, more than one-half were against this as it may 'make financial statements bulky', 'create confusion' and will 'not facilitate decision-making by stakeholders'. The study found existence of a significant environmental reporting expectation gap in India.

'Usefulness in decision-making' was considered as the most important characteristic of good environmental statements by majority of the CAs and the executives. For this, they felt that the information should be relevant and objective. They were of the view that benefits of environmental information to the potential users are greater than the cost of providing it. The environmental statements should be prepared keeping in mind users' needs. 'Society at large' was considered to be the most relevant user of environmental information by the CAs. 'Government', 'environmental organisations' and 'owners' were other relevant users, according to them. However, the executives found 'government' to be the most important user of this information because these statements were mainly prepared by their companies to comply with the legal requirements. It helped them in reducing pressures from regulatory authorities.

7.5 Environmental Audit

A large majority of the CAs were of the opinion that environmental statements prepared by a company should be verified by an auditor to 'increase credibility of information provided' in them. They felt that this audit will also help a company in complying with environmental laws. They, in general, favoured appointment of a team, mainly comprising of external environmental auditors to do this task. 'Chartered accountants', 'engineers' and 'cost accountants' may also form a part of this team to handle specific tasks, according to them.

A scrutiny of actual practices followed by the companies revealed that most of them prepared statutorily required environmental statements, while about 39 percent also made these reports voluntarily. 48 percent of the companies (which prepared environmental statements) got these statements verified by an auditor. The task of environmental information audit was generally given to a team of auditors. About four-fifths of the companies appointed external environmental auditors for this purpose, while some companies assigned specific tasks to engineers and CAs.

7.6 Analysis of Annual Reports

An examination of environmental disclosure practices of the selected Indian companies was made by analysing annual reports of 91 companies for 3 consecutive years from 1999-2000 to 2001-2002. For this purpose, an 'Index of environmental disclosure' consisting of 23 items of environmental information that a company can disclose, was constructed. The CAs were requested to evaluate each of these items independently and assign them ranks on the basis of their importance and usefulness in making sound decisions and evaluating environmental performance of a company. The CAs considered all items given in the 'Index' as of some importance. Three items were considered as 'most important' by them. These were: 'prevention and repair of environmental damage', statutorily required measures taken for 'pollution control' and 'conservation of natural resources'. In general, they were of the view that a company must disclose some information about all these items as these are useful for various stakehoiders in making relevant investment and other decisions.

An analysis of environmental disclosure practices of the companies with the help of this 'Index' revealed that the companies failed to provide adequate environmental disclosures in the annual reports. In the absence of any accounting standard or guideline, both quantity and quality of disclosure was quite unsatisfactory. Moreover, there was no consistency in these disclosures over the years. The common practice followed by these companies (of all industries and sectors) over the years was to provide descriptive information and itemise the same in the Director's Report. Reporting of quantitative/financial information on environment was almost negligible. Most of the times, the disclosures were quite general in nature and in little specific details. Majority of the companies provided only statutorily required information on 'conservation of energy' in the annexure to the Director's Report. Even when voluntary reporting was there, it was generally in few lines only. The majority of such disclosures could be characterised as statements of good intentions, legal compliance and selective self-congratulatory assertions. These disclosures appeared to be made mainly for image building, improving public relations and reducing legal pressures. However, they generally lacked in information content and transparency. Thus, most of the companies provided mainly qualitative, statutorily required and positive information on environment.

7.7 Multivariate Analysis

An attempt was made in the study to statistically test the combined impact of selected company- and industry-related variables on environmental reporting by using multiple regression technique. It was observed that all the selected variables together explained the variation in environmental disclosure practices of the companies significantly. Environmental reporting by the public sector companies was found to be significantly higher than those operating in the private sector. Moreover, the 'most polluting' companies were found to be communicating significantly more information on environment than the 'less polluting' concerns. Among quantitative variables, 'environmental performance' was the most important explanatory variable. Companies with better environmental performance provided detailed information on environment than poor performers. The results of the study provided evidence in support of the hypotheses related to influence of some variables like 'size', 'profitability', 'sector', 'industry' and 'environmental performance' on environmental disclosure practices. However, the hypotheses related to impact of 'association with large business houses', 'foreign association', 'debt-equity ratio' and 'exports' on environmental reporting were not substantiated by the findings of the multivariate analysis.

During last one and a half decade, the green accounting and reporting has received increasing attention from the academicians, researchers, companies and regulatory authorities. Though some of the countries like Denmark, the Netherlands, Spain, USA and Australia have implemented mandatory reporting on environmental matters, no accounting body has yet issued any specific guideline or standard on EAR. In fact, to improve the quality of EAR, there is a need to develop a conceptual framework of EAR, and subsequently, a standard based on it containing objectives, general assumptions, qualitative characteristics and guidelines for companies.

The findings of the study reveal that accountants in India are well aware of the fact that environmental issues will affect business and industry in the near future. They are convinced by the need for accounting of environmental information. Despite this awareness, there is an absence of external environmental accounting. The companies in India do not have a proper environmental accounting system to determine environment related costs, benefits, assets and liabilities.

No significant difference was found between the views of the users and the preparers on various issues raised in the study. Both the CAs and the executives found usefulness of environmental information in strategic decision-making. They favoured mandatory environmental reporting in India. They felt that benefits of providing environmental information to the relevant users are more than cost of providing this information. But when actual environmental reporting practices were analyzed, significant differences were found in expectations of the users and actual practices of the companies. Though non-financial/descriptive disclosures in the annual reports increased overtime, yet the quantitative and financial environmental reporting was negligible. Without any strict accounting pronouncements from the ICAI and disclosure norms by the regulatory authorities, the companies generally provided only statutorily required, qualitative, and positive information on environment. These disclosures lacked in quality and consistency. The evidence is consistent with the existence of an EAR expectation gap within India. It can, thus, be concluded that there is a low level of EAR activity in India. The accounting profession in India has yet to respond to the issue of environmentalism and help the companies in dealing with environmental matters in the books of accounts.

7.8 Recommendations

The present study has covered three areas namely, environmental accounting, reporting, and auditing. Keeping in view the observations of the study, suggestions have been made in each of these three areas. The recommendations are for the benefit of professional accounting bodies, government, regulatory authorities, professional accountants, companies, and various stakeholders.

7.8.1 Environmental Accounting: Environmental accounting at the corporate level includes preparation of environmental accounts for external reporting purposes as well as use of this information in internal decision-making. It was observed that in the absence of any regulatory framework and accounting guidelines on the issue, the companies in India find it difficult to incorporate environment related information formally in the books of accounts. Though the companies have started considering these issues for managerial decision-making, most of them do not prepare environmental accounts for external reporting. The following recommendations have been made to improve the use of environmental accounting system in India:

The study reinforces a great need for development of a separate conceptual framework and a standard on EAR to help companies in properly identifying, measuring, recognising and disclosing relevant environmental costs, benefits, assets, liabilities and contingencies. Hence, it is recommended that professional accounting bodies at national and international levels should develop a separate conceptual framework on EAR (on the lines prescribed by FEE) specifying objectives, general assumptions, qualitative characteristics and guidelines for the companies. A standard should also be issued giving basis on which environmental costs should be bifurcated from other costs, basis for expensing and capitalisation of environmental costs, depreciation of environmental assets, recognition of environmental benefits and measurement and recognition of environmental liabilities and contingencies. Some guidelines should also be provided for treatment of environmental costs relating to prior period in the books of accounts and on recovery and impairment of environmental assets. Existing accounting standards can also be modified to take care of environmental issues. The ICAI should also develop a separate standard on Carbon (Emission and Sequestration) Accounting (CES Accounting) which needs to be consistent with the Intergovernmental Panel on Climate Change (IPCC) principles. In addition, the measurement of externalities, tax treatment of environmental activities and verification of an entity's environmental accounts by independent duly qualified assurors are some other issues to be considered by the ICAI.

With growing legal and social awareness, there has been a significant increase in environmental costs and liabilities, to the extent that they have become a significant part of total costs and liabilities. Concealing them or their improper disclosure might expose interested stakeholders to unknown levels of risk. Hence, significant environmental costs and liabilities should be properly measured and reflected in the financial reports. It might not be possible for a company to account for external effects of environmental pollution and degradation within the current accounting framework. But the enterprises should take initiative to track internal environmental costs and benefits to products and processes. They should try to determine full extent of their environmental costs and integrate them into decision-making.

It has been observed that the companies generally emphasize ever increasing environmental costs, but environmental benefits (even significant financial benefits) from incurring these costs are rarely discussed. The study recommends proper measurement and separate recognition of significant financial benefits of being eco-efficient in the financial statements. In addition, social/ intangible costs and benefits can be disclosed in the footnotes, annexure or separate environmental statements. The preparers of financial statements should be acquainted with the benefits of adopting an environmental accounting system. The significance of environmental information in making strategic business decisions, in complying with environmental laws and other benefits should be popularised among CAs and company officials.

7.8.2 Environmental Reporting: The growing social, legal and judicial pressures have forced companies to disclose more and more information on environment in their annual reports. At present environmental reporting is not mandatory in India (except on energy conservation). Though some companies have started taking initiatives in this regard, reporting on these issues is still very limited in India. The following suggestions have been made to improve environmental reporting by the corporate sector:

The findings of the study suggest that the main reason for

non-disclosure of environmental information by the Indian companies is its voluntary nature. It is strongly recommended that environmental reporting should be made mandatory in India. As a first step, Indian Companies Act and other related statutes should be amended suitably requiring the companies to disclose information on some important items like measures taken for pollution control and conservation of natural resources, prevention and repair of environmental damage and legal compliance in the annual reports. Later on, some other items can be selected for compulsory disclosure in the annual reports in different phases.

In the absence of any accounting standard or guideline, identification, measurement and recognition of environmental costs, benefits, assets, and liabilities is, to some extent, judgemental. Till some concrete initiatives are undertaken by the Government and other regulatory bodies on this issue, companies should be required to provide a brief description of significant environmental accounting policies in the annual reports. It would reduce confusion in the mind of the users of these statements.

Companies in India (particularly those in the most polluting industries) should be asked to submit detailed environmental information to the government regarding emission of specific toxic chemicals, pollutants, effluents, damage to the environment and community health. They should also inform about the steps taken to reduce these externalities. In USA, Toxic Release Inventory (TRI) is maintained by Environmental Protection Agency (EPA). In Australia, National Pollutant Inventory (NPI) is recorded. On the similar lines, some records should be kept by Ministry of Environment and Forests (MoEF) in India, so that necessary steps to protect environment can be taken in time.

SEBI may require listed companies to disclose information on some important items like compliance with environmental laws, pending judicial proceedings arising under environmental laws and significant environmental liabilities on a prescribed form. For this purpose, a clause in the listing agreement (like clause 49 for Corporate Governance) may be inserted for strict compliance.

Companies must disclose all types of relevant environmental information like qualitative information (e.g., environmental objectives, policies, compliances); quantitative information (e.g., emissions, toxic releases, pollution levels, consumption of scarce natural resources) and financial information (e.g., significant environmental costs, benefits, assets and liabilities). A company can make narrative disclosures or present information by way of tables or graphs. In addition to positive information, significant negative developments like legal proceedings, major accidents or significant environmental degradations should also be disclosed. Needs of relevant users should be kept in mind while disclosing environmental information and deciding about the mode of disclosure.

In the annual reports, environmental disclosures can be made at different locations. In fact, different types of environmental information should be disclosed at different suitable places. For example, company's commitment for environmental protection, broad objectives and policies can be shown in Chairman's or CEO's statement. Other qualitative disclosures (e.g., information about company's environmental performance, legal compliance, establishment or existence of EMS, ISO 14000 certificates, steps taken to control pollution and protect environment) can be made in the Director's Report or its annexure. Significant financial information on environment (e.g., significant environmental costs, benefits, assets and liabilities) must be disclosed in the main body of the financial statements. In addition, environmental contingencies, provisions, significant environmental accounting policies can be shown in the footnotes or annexure to financial statements.

The environment is only one of the strategic issues that need to be addressed in the annual reports. Keeping in mind the limited space available, it may not be possible for companies to provide detailed information on environment in these reports. But environmental information should at least be given equal disclosure opportunities along with other strategic issues. When restricted by space in the main financial statements, the users should be referred to for further details in the annexure or notes to these statements. Further, to avoid any confusion for the users of the financial statements, the company may disclose detailed environmental information through stand-alone corporate environmental reports (CERs) or websites for the benefit of various stakeholders. It will also help a company in highlighting importance of these issues.

7.8.3 Environmental Audit: To increase credibility and reliability of environmental information supplied to various interested stakeholders, it is necessary that it should be verified by some independent auditors. The following recommendations may help in ensuring accuracy and authenticity of information provided in these reports:

In major polluting industries, environmental audit should be made mandatory. This can be extended to other industries, after the audit systems in the polluting industries are well documented. An Institute of Environmental Accountants and Auditors of India (IEAAI) may be established as a statutory body under an Act of Parliament to regulate the profession of environmental accountants and auditors. This institute will help in the field of environmental education, professional development as well as maintenance of high sustainability accounting, auditing and ethical standards. In the meantime, the ICAI should issue some guidelines or standards for verification of environmental information.

In the absence of any accounting standard or legal requirements on the issue, companies should voluntarily provide environmental information verified by external auditors. They should appoint a team of external auditors to audit environmental statements and should entrust this task to some qualified environmental auditors. Professionals from other areas like chartered accountants, cost and work accountants, and engineers can also be appointed to handle specific tasks. Environmental audit report should be published along with annual financial statements or separate environmental statements to reduce the communication gap between the public and the industry. A copy of this report should be supplied to Ministry of Environment and Forests (Government of India) by companies in polluting industries.

In general, regulatory authorities (like SEBI and RBI) should encourage companies to spend more on environment by giving tax benefits, exemptions and preferential treatment to eco-efficient concerns in relation to matters like registration, listing or clearances. Professional bodies like ICWAI and ICAI should encourage companies to become environment friendly. Companies and professional accountants should be made aware of the benefits of being eco-conscious, e.g., significant social and financial benefits in using eco-efficient processes and technologies. Environmental accounting and auditing should be included in the curriculum for production management, environmental engineering, industrial management and specialised professional courses. Investors, consumers, employees and other stakeholders should also emphasise on the environmental performance of the company while dealing with a company.

7.8.4 Future Directions: As mentioned earlier, the present study is global in nature covering various aspects relating to EAR. It is based on primary as well as secondary data.

Given the tremendous importance of environmental issues for social welfare, further research is desired to gain more insight into the issues raised in the study. Separate in-depth studies covering various issues in each of the areas– environmental accounting, environmental reporting, and environmental audit can be undertaken for this purpose. Moreover, the opinions of other stakeholders like investors, creditors, customers and workers can be obtained on various issues relating to accounting and disclosure of information on environment.

Though findings of the study are quite comprehensive, relatively small sample size, at times, inhibits generalising the results. Therefore, another study with a larger group can be conducted to generalise the results of the present study.

The present work is limited to large manufacturing companies situated in India only. Another research covering environmental issues in non-manufacturing sectors (like banks, hotels and educational institutions) can be undertaken. The scope of the study can further be extended by comparing the EAR practices of Indian and overseas companies using a common questionnaire duly adjusted for cultural, social and political influences. In addition, focussing on environmental practice of small and medium enterprises (SMEs) can make significant contributions to the existing knowledge on the subject. A comparative analysis of the environmental reporting systems of the public and private limited companies can also help the diversified stakeholders in making informed decisions.

This research has examined environmental information disclosure in annual reports only. An analysis of disclosure of information on environment by companies through other mediums like Internet, press releases and separate corporate environmental reports (CERs) can be conducted.

Last but not the least, a study developing ways to recognise and measure social costs and benefits can be another pioneering research area.

7.9 A Final Word

Environmental issues have increased in importance in the past two decades. The concepts of EAR are now well established and are continuing to develop rapidly. In 21st century, there is an increasing acceptance to the need for EAR and sustainability accounting. That this trend promises to continue is reflected in the increasing financial attention firms are giving to these issues. But the way in which most businesses accept or use EAR suggests that they have little understanding of what EAR is and how to implement it. One basic reason for this could be lack of accounting standards and guidelines on the issue. This book has been written with a very specific objective of providing a basic introduction to the concepts of EAR in India. It provides various theoretical considerations involved in EAR, regulatory environment in different countries, a comprehensive review of existing literature and evidences on current practice.

This introduction to the concepts and the legal requirements of EAR in different countries together with evidences on current practices would help companies and practitioners who want to begin the process of addressing environmental issues in their work. They will realize the need to be aware of environmental issues and to consider their influence upon both internal and external reporting. The evolution of the appropriate financial treatment for all these issues provides a challenge to the accounting profession in India. It must be understood that in responding to this challenge posed by the environment, all aspects of accounting including financial reporting, auditing, management accounting and taxation may have to be modified. Research into issues such as carbon pricing, life-cycle costing, environmental taxes, sustainability accounting, and carbon efficiency management, would perhaps be avenues by which the accounting profession can respond to the issue of environmentalism and can play its part in reducing global warming.

Appendices

Appendix I: Mandatory Environmental Reporting Schemes in Selected European Countries

Denmark	Netherlands	Norway	Sweden
Scheme			
Green Accounts	Environmental Reporting	Environmental Information in Directors statement in annual reports	Environmental Information in Annual Report
Legal Frame work			
1. Act No. 403 of June 14, 1995 Amending the Environmental Protection Act 2. Detailed rules in the Statutory order from the Ministry of the Environment and Energy No. 975 of December 13,1995	Act of Parliament dated 10 April 1997 to extend the Environmental Management Act and incorporate a new Title 12.1 entitled 'Environmental Reporting' 2. Environmental Reporting Decree, enforce from 1 Jan. 1999.	The Norwegian Accounting Act (1.1.1999) – section directors report	Act of Parliament dated 1 January 1999 to extend the Law of Account (1995:1554) and incorporate a new part to Ch. 6 (Directors Report / Annual report) (For valtnings berattels e)
Description			
The scheme sets rules for listed companies to prepare annual green accounts. The statement of accounts shall indicate the significant consumption of energy, water and raw materials and the type/quantity of pollutants, forming part of the production process, which are discharged from the enterprise to air, water and soil or form part of the products and waste.	Establishes a regime by which establishments licensed by the province and which have a substantial environmental impact can be required to produce each year an environmental report for the authorities (government report) and an environmental report for the general public (public report)	Information about conditions which may affect external environment. An account must be given of matters relating to the enterprise, including its resource used in production and products, which contributes to an impact on the external environment and of the measures which have been implemented or are being planned to prevent or reduce negative impact on the environment.	The scheme sets for companies obliged to report and seek permit according to the Environmental code to report the enterprise's effects on the surrounding environment caused by its production processes and whether the effects have direct or indirect significance to its financial development in the annual report

Detailed Rules: Specific guidelines on information content & format			
Yes (Subsection 5 of the statutory order)	Yes for the government report (Annex 2 of the decree) For the public report it will be free-format	The Accounting Act specifies several detailed demands about information on environment that intended to be answered in the Directors Report (Annual report)	Yes The Accounting Act details demands about what type of information
Scope: Affected Companies			
All companies fall under the approval obligation of the Environmental Protection Act Number of companies under this category = 3000, currently 1200 are reporting Reporting is at site level	Establishment of license under the Environmental Management Act. Currently 250 companies will be designated reporting companies Reporting is at site level	All companies under companies law Reporting is at company level	All companies that are obliged to report and seek permits according to Ch9, pg. 6 of The Swedish Environmental Code. Currently affects some 10,000 enterprises Reporting is at company and site levels. Companies with operations outside Sweden need not to report about these sites.
To whom the accounts are submitted?			
1. Local Supervising authority 2. Danish Commerce and Companies Agency Public can access the reports.	The act requires the company to send two copies of the government report both to the province administrative body granting the license, and to the regional department of public works and water management.	Accounts are submitted to the national register of company data, administered under the Ministry of Finance. Public can access this register	Registration Authorities : patent & registreringsverket PRV9 Public can access this information through the authorities.

Verification/audit requirements			
No obligation for a third party audit however the rules state that in case there is an internal or external audit, the name and auditor's declaration shall be stated in the accounts	The decree stipulates that there should be an independent verification of the report but it has not been implemented	Annual reports need to be audited however this does not apply to information in the directors report	Annual reports need to be audited.
Administration and Enforcement			
The Danish Commerce and Companies Agency administer the acceptance and publication of accounts on its information system. Local supervisory authorities or the Environmental Protection can decide whether the accounts can be published or not.	Companies are obliged submit the government report to the licensing authority. Coordinating competent authority will administer a centralized database where the information is shared among the various agencies involved in monitoring and enforcement.	Accounts are submitted to the national register of company data, administered under the Ministry of Finance	Accounts are submitted to the Patent and Registration office (PRV).
Exemptions			
If the company is implementing EMAS, the annual environmental statement can be considered a substitute if validated by an accredited auditor	If the company issues an annual environmental statement based on EMAS, the company is exempted from making a public report.	No exemption. All companies registered in Norway shall try to include environmental information in the annual reports	Exemption is given to sites outside Sweden.
Fines			
Yes	Yes. Failure to comply with the obligations may give rise to the imposition of a performance bond as an administrative sanction.	None	Yes

Evaluation			
Following a transition period of three years, the number of reporting companies is expected to increase from 1200 to 3000. New guidelines are also being introduced	The scheme is to be evaluated after 3 years of operation (2002)	In an unofficial evaluation conducted by GRIP [www.grip.no], they conclude that the performance of large companies in regard to this requirement has been very poor. However the law is expected to be revised in year 2002.	NA

Note: Emtairah T., "Corporate Environmental Reporting: Review of Policy Action in Europe", IIIEE Report, Lund, Feb. 2002.
Source: http://www.enviroreporting.Com/others/cer-europe.pdf.

Appendix II: List of 17 Most Polluting Industries in India

S.N	Industry
1	Aluminium Smelter
2	Caustic Soda
3	Cement
4	Copper Smelter
5	Distilleries
6	Dyes & Dye Intermediates
7	Fertilizer
8	Integrated Iron & Steel
9	Tanneries
10	Pesticides
11	Petrochemicals
12	Drugs & Pharmaceuticals
13	Pulp & Paper
14	Oil Refineries
15	Sugar
16	Thermal Power Plants
17	Zinc Smelter

Major Environmental Laws in India

- Water (Prevention and Control of Pollution) Act, 1974;
- Air (Prevention and Control of Pollution) Act, 1981,
- Cess Act, 1977,
- Environment (Protection) Act, 1986 and Rules thereunder
- Public Liability Insurance Act, 1981.
- National Environmental Tribunal Act, 1995
- National Environment Appellate Authority Act, 1997

Appendix III: Carbon Trading

In modern times, the burning of fossil fuels like coal, oil and natural gas in industrial production, rising vehicular pollution levels combined with accelerated land clearance and deforestation has led to unprecedented levels of greenhouse gas (GHG) emissions. The concentrations of greenhouse gases in the atmosphere have risen dramatically leading to an out-of-balance greenhouse effect. Most scientists say that as concentrations of these gases continue to rise, there will be a general and very rapid warming of the world's climate. The effects of such warming may range from dire to mild, i.e., from widespread ecological changes in agricultural production and rising sea levels, to moderate changes in regional climates. It is predicted that if emissions of greenhouse gases continue to increase at the current rate, around 50 percent of the species on earth may become extinct during this century. Global warming will also lead to destabilization of the massive ice sheets in Greenland and Antarctica. With rise of sea levels, the world will lose coastal cities like Mumbai, Tokyo, New York, displacing millions of people. Thus, Global Warming is not just an environmental or health concern- it is a matter of human survival.

People have become increasingly concerned about the effects of global warming and resulting Global Climate Change (GCC). Governments and supra-national bodies have sought to respond to GCC in a variety of ways like by introducing carbon trading or imposing environmental taxes. The need for a reduction in carbon emissions was debated at the United Nations Conference on Environment & Development (The Earth Summit) in Rio de Janeiro in 1992, resulting in the adoption of United Nations Framework Convention on Climate Change (UNFCCC) which is designed to impose limits on greenhouse gas emissions and thus minimize the adverse effects of climate change. Subsequently an international treaty on environment was entered into known as Kyoto Protocol. The Kyoto protocol is a 1997 international treaty which came into force in 2005. It is the first step towards stabilizing global emissions of carbon dioxide. It binds most of the developed nations to a cap and trade system for 6 GHGs. The main objective of the Protocol is achieving quantified emission limitations through specific policies and measures to minimize the adverse effects of climate change.

Under the UNFCCC, countries are permitted to use a trading system to help meet their emissions targets. The process of buying and selling of carbon credits is known as Carbon Trading (sometimes called cap and trade system). It is an administrative approach used to control pollution by providing economic incentives for achieving reductions in the emission of pollutants. The overall goal of an emissions trading plan is to reduce emissions of harmful GHGs. Trading of carbon credits occurs through certified emissions reductions (CERs). A CER is a certificate issued by the authorized body (CDM Executive Board) to the approved projects that have reduced GHG emission in a calendar year. For example, the emission of GHGs by an approved project shall be continuously measured for a period of one calendar year and the company concerned is entitled to receive 100 CERs if it has emitted 100 tons of GHGs lesser than its maximum allocated GHG emission limit. In principle, a country may allocate permits to individual companies for the emission of a certain quantity of greenhouse gases. If permits are only issued to a level equal to or below the assigned amount, then a country should meet its Kyoto commitment (assuming that the measures of its emissions are accurate). If a country is incapable of meeting its target, it can buy permits from countries that are under their targets. Similarly, companies within a country that prove more able to reduce their emissions are allowed to 'trade' excess permits to other, more polluting enterprises. [1]

The Kyoto protocol prescribes three mechanisms which will help in reducing GHG emission, namely Clean

Development Mechanism (CDM), Joint Implementation (JI) and International Emission Trading (IET). Under CDM, industrial organizations of developed countries can achieve the compliance by buying CERs from GHG reducing projects approved by UNFCCC set-up in developing countries. Under JI, a developed country with relatively high costs of domestic greenhouse reduction can set up a project in any developing country with relatively low costs for greenhouse gas reduction. Under IET, countries with shortage in carbon credits can buy from the countries with surplus carbon credits.

Carbon markets have the effect of putting a price on what was until very recently free and this change is likely to have financial consequences for firms in the longer term. As regards accounting issues, there are problems associated with the valuation of pollution allowances and their identification as assets (and the liabilities that arise if companies pollute beyond allowed levels). Till date, accounting standard setters have not issued clear guidelines on accounting of carbon trading. For example, it is not very clear that once allowances are issued, whether a company should recognize them as a new asset on the balance sheet. As actual emissions occur, whether a liability should be recognized and whether changes in the market price of allowance (i.e., gains and losses on allowances) are to be recognized in the profit and loss account. Companies will also need to consider issues such as fair value accounting and impairment of assets. It has been left to organizations outside the accounting profession to develop CES accounting, measurement, reporting frameworks. There are only few detailed approaches and metrics available on the issue and many of them are incompatible with each other. This lack of consistency has then resulted in almost no development in assurance standards, from within or outside the accounting profession. Companies to date have very little guidance from accounting and assurance standard setters as to the treatment of carbon related intangible assets.

The environment policy-related carbon trading has entered

India also with many companies in India have started establishing systems for reducing carbon dioxide emissions in the atmosphere. The Multi Commodity Exchange of India Ltd entered into an alliance with the Chicago Climate Exchange to introduce carbon credit trading in India in 2005. This association has integrated Indian markets with their global counterparts to cover risks associated with future trading of carbon credits and ensuring best prices. India's carbon credits' trading is expected to reach $100 billion by 2010. In 2007, a total of 160 new projects were registered with UNFCCC. The accounting issues arising due to carbon trading must be answered by accountants in India also one of the most important is the accounting treatment of emission certificates which are intangible assets like patents or copyrights. Valuation of these certificates at cost or at market price, amortization of these certificates, and measurement and recognition of related costs, benefits, assets and liabilities are many questions which are required to be decided by accountants. Though emission trading has come to India (within and outside the Kyoto Protocol), but the accounting standard setters have yet to come up with a standard or guideline to account for such activity. The Institute of Chartered Accountants of India (ICAI) must work on accounting norms for carbon credits.

Carbon trading has given rise to many taxation issues also like whether CERs can be classified as goods or services for indirect tax purposes. Consequently, whether CERs would attract VAT (levied on sale of goods) or service tax (levied on rendering of services) and whether the sale of CERs to overseas buyers would qualify as export of goods or services is also not clear. To determine whether VAT or service tax would be applicable, it is relevant to determine whether CERs are classifiable as goods or services. At present, indirect tax legislation does not provide guidelines for such treatment and jurisprudence on this issue is yet to evolve. Lack of clarity on the treatment of CERs for indirect tax purposes exposes companies engaged in projects generating carbon credits to the threat of significant tax demands.

In nutshell, public concern the world over has resulted in building mechanisms such as the Intergovernmental Panel on Climate Change (IPCC) and the Kyoto Protocol for Greenhouse gas (GHG) reduction. Many issues have emerged due to introduction of carbon trading. Governments need to consider carbon regulation issues such as rationing or taxing of net-CO_2 emitting entities (organizations and individuals) and providing credit allowances or tax-breaks for net-CO_2 absorbing entities. On the other hand, business entities need to consider issues such as trading in carbon allowances (or permits), investment in low- CO_2 emission technologies, counting the costs of carbon regulatory compliance and passing on the increased cost of carbon regulation to consumers through higher prices. These decisions and their consequences will impact the accounting profession significantly. Unfortunately, in terms of financial reporting, the current financial accounting framework in India appears to be ill equipped to provide the information required by companies to meet the challenge of global warming. This is mainly because accounting information systems based on the accounting equation are not designed to cope with non-monetary measures such as CO_2 sources and sinks. As such, despite the fact that emissions trading has come to India with big bang, the accounting standard setters have yet to come up with an acceptable standard to account for such activity.

Note

1. According to a survey by PricewaterhouseCoopers (PwC) and the International Emissions Trading Association (IETA), in the absence of clear international guidance on the issue companies are employing a variety of different accounting practices for carbon emissions. The survey found that companies covered by the EU Emissions Trading Scheme (ETS) had applied 15 distinctive approaches when accounting for EU allowances (EUAs), although there were six main approaches. Such

diversity in accounting practice led to companies with similar emissions profiles providing very different results on the balance sheet, It is difficult to compare the business performance when accounting treatments are not clear or vary so greatly.

Sources

Bebbington, J. and Larrinaga, C., "Carbon Trading: Accounting and Reporting Issues", European Accounting Review, Vol. 17, Issue 4, Dec.2008, pp.697-717.

Palash Goswami, "Accounting and Reporting of Carbon Credits and Their Trading", The Accounting World, Oct. 2008.

Ratnatunga, J., "An Inconvenient Truth about Accounting", Journal of Applied Management Accounting Research, 5(1), 2007.

Appendix IV: Some Sources for EAR Guidelines

ACBE	(UK Government's) Advisory Committee on Business and the Environment	Environment Reporting and the Financial Sector- An Approach to Good Practice(1997)
ACCA	Association of Chartered Certified Accountants (UK)	Guide to Environment and Energy Reporting and Accounting (1997)
CBI	Confederation of British Industry (UK)	Introducing Environmental Reporting Guidelines for Business (1993)
CIA	Chemical Industries Association (UK)	Reporting to your Local Community(1995)
CICA	Canadian Institute of Chartered Accountants (Canada)	Reporting on Environmental Performance (1994)
D&T	Deloitte & Touche (Denmark)	Assessor's Manual for the Analysis and Evaluation of Corporate Environmental Reporting (1996)
EFFAS	European Federation of Financial Analysis Societies	Eco-Efficiency and Financial Analysis- The Financial Analyst's Views (1996)
EMAS	EC Eco-Management and Audit Scheme	Requirements for the Environmental Statement (1998 revision)
FEEM	Fondazione ENI Enrico Mattei (Italy)	Company Environmental Reports-Guidelines for Preparation (1995)
GEMI	Global Environmental Management Initiative (USA)	Environmental Reporting in a Total Quality Management Framework (1994)
GRI	Global Reporting Initiative	Sustainability Reporting Guidelines (1992/2000)
		Business, Accountancy and Environment: A Policy and Research Agenda (1992)

ICAEW	Institute of Chartered Accountants in England and Wales (UK)	Environmental Issues in Financial Reporting (1996)
ICC	International Chambers of Commerce	The Business Charter for Sustainable Development (1991)
ISAR	United Nations	Conclusions on Accounting and Reporting by Transnational Corporations (annual)
100 Group	The 100 Group of Financial Directors (UK)	Environmental Reporting : The Hundred Group Statement of Good Practice (1991)
PERI	Public Environmental Reporting Initiative (USA)	The PERI Guidelines (1992)
UNEP	United Nations Environment Programme and Sustainability	Engaging Stakeholders; 2nd International Progress Report on Company Environmental Reporting (1996)
WICE	World Industry Council for the Environment (now art of WBCSD)	Environmental reporting – A Management Guide (1994)

Source: FEE Discussion Paper, "Towards a Generally Accepted Framework for Environmental Reporting", 2000.

Some Descriptions

Name of guideline	**Description**
The Sustainability Reporting Guidelines of the Global Reporting Initiative (GRI) 2006 (G3)	The Global Reporting Initiative's Sustainability Reporting Guidelines represent the first global framework for comprehensive sustainability reporting, encompassing the "triple bottom line" of economic, environmental, and social issues. The Guidelines bring reporters and report users alike one major step closer to elevating sustainability reporting to a level consistent with financial reporting

UNCTAD Report: "Environmental Financial Accounting and Reporting at the Corporate level	The Intergovernmental Working Group of Experts on International Standards of Accounting and Reporting (ISAR) of UNCTAD published in 1998 this report. The first chapter is an interim statement of best practice guidance for accounting and reporting for environmental costs and liabilities. The purpose is to provide assistance to enterprises, regulators and standard-setting bodies in determining what is considered best practice in accounting for environmental transactions and events in the financial statements and associated notes.
The INEM Sustainability Reporting Guide: A Manual on Practical and Convincing Communication for Future-Oriented Companies	The guide (October 2001) gives advice on how to draw up an effective sustainability report. Focusing on the economic interests of companies, as well as on reliable and convincing communication, it informs about target groups and their interests in information, outlines the principles of sustainability reporting and describes the elements of the sustainability report. It also provides in detail numerous practical examples and illustrations for presentation of performance data and design of sustainability reports.
CMA Management Accounting Guideline: "Writing and Evaluating Sustainable Development and Environmental Reports" (exposure draft	The objective of the guideline issued in 1998 by The Society of Management Accountants of Canada (CMA) is to provide instruction on how corporate commitment to sustainable development (SD) and environmental stewardship practices can be most effectively communicated through SD and environmental reports. It also assists management accountants and other stakeholders in the assessment of these reports.

ACBE-guidelines "Environmental Reporting and the Financial Sector: An Approach to Good Practice	In February 1997 the Advisory Committee on Business and the Environment (ACBE) in the UK issued guidelines directed at all businesses encouraging more widespread use of environmental reporting and set out good practice for reporting the financial aspects of environmental performance
The CEFIC Responsible Care - Health, Safety and Environmental reporting guidelines	The European Chemical Industry Council (CEFIC) has published guidelines in 1998 for the European chemical companies. Their previous guidelines of 1993 are also well-known in other sectors
The WICE guidelines	The World Industry Council for the Environment (WICE) is a global coalition of enterprises initiated in 1993 by the International Chamber of Commerce (ICC). In 1995 they published their guidelines for voluntary reporting
The ACCA Guide to environment and energy reporting 1997	In 1998 the accountants' organization ACCA (UK) published this guide which contains step-by-step advice on how and when to report environmental issues to stakeholders. It features a best practice checklist, an introduction to environmental reporting issues, a guide to reporting on energy use and has sections on report development.

Source: Sustainability-Reports.com

Appendix V: Testing for Superfluous Variables and Multicollinearity

Table 1: Testing for Superfluous Variables (Type I_a equations)

S.N	Constant	D_S	D_{MP}	D_{FA}	D_{BH}
1	–.0482	.110 (2.211)**	.0758 (2.193)**	.0491 (1.083)	.0461 (1.282)
2	–.0076	–	.0977 (2.886)***	.024 (.536)	.01112 (.338)
3	–.0465	.141 (2.901)***	–	.0432 (.934)	.0549 (1.502)
4	–.0249	.0964 (2.002)**	.0736 (2.131)**	–	.0353 (1.022)
5	–.0259	.0819 (1.827)*	.0807 (2.342)**	.0331 (.757)	–
6	–.0713	.145 (3.052)***	.0781 (2.222)	.0535 (1.160)	.0477 (1.304)
7	–.0462	.0978 (1.96)*	.0776 (2.219)**	.0512 (1.117)	.0468 (1.287)
8	–.0405	.104 (2.15)**	.0780 (2.28)**	.0503 (1.116)	.0433 (1.223)
9	–.0631	.119 (2.39)***	.0648 (1.91)**	.0531 (1.17)	.0409 (1.14)
10	.108	.0937 (1.787)*	.103 (2.88)***	.028 (.583)	.0459 (1.21)

Contd...

S.N	S	P	E	D/E	EP
1	.000003 (2.031)**	.00099 (1.737)*	.00063 (.558)	–.0144 (–1.406)	.01668 (3.33)***
2	.000004 (2.915)***	.00081 (1.408)	.00009 (.082)	–.0172 (–1.658)	.0156 (3.058)***
3	.000003 (2.058)**	.00102 (1.763)*	.00092 (.795)	–.00934 (–.915)	.0193 (2.901)***
4	.000003 (2.083)**	.0010 (1.765)*	.00069 (.610)	–.0151 (–1.477)	.0159 (3.206)***
5	.000003 (2.05)**	.00099 (1.745)*	.00043 (.385)	–.0131 (–1.279)	.0167 (3.316)***
6	–	.00113	.00052	–.0120	.0189

		(1.975)*	(.454)	(–1.161)	(3.81)***
7	.000003 (2.24)**	–	.00079 (.686)	–.0161 (–1.56)	.0168 (3.31)***
8	.000003 (2.015)**	.001 (1.79)*	–	–.014 (–1.41)	.0165 (3.32)***
9	.000003 (1.87)*	.0011 (1.87)*	.0006 (.554)	–	.0171 (3.39)***
10	.000004 (2.68)***	.001 (1.68)*	.0004 (.35)	–.0163 (–1.50)	–

Contd...

S.N	R^2	R^2	F-Value	Sig. Level
1	.401	.334	6.018	.000
2	.365	.303	5.881	.000
3	.365	.303	5.896	.000
4	.392	.334	6.610	.000
5	.389	.329	6.514	.000
6	.370	.309	6.026	.000
7	.378	.318	6.240	.000
8	.398	.340	6.789	.000
9	.386	.326	6.447	.000
10	.319	.252	4.794	.000

Note: 1. T-Values are given in the parentheses; 2. *Significant at 10% level of significance; **Significant at 5% level of significance; *** Significant at 1% level of significance.

Table 2: Testing for Superfluous Variables (Type I_b equations)

S.N	Constant	D_S	D_{MP}	D_{FA}	D_{BH}
1	–.0544	.127 (2.49)**	.0887 (2.540)**	.0477 (1.04)	.0457 (1.25)
2	–.0041	–	.114 (3.33)***	.0187 (.409)	.0058 (.172)
3	–.0565	.165 (3.29)***	–	.0395 (.836)	.0562 (1.498)
4	–.0312	.114 (2.30)**	.0861 (2.47)**	–	.0352 (1.00)
5	–.032	.0991 (2.157)**	.0936 (2.69)***	.0318 (.718)	–
6	–.0866	.163 (3.37)***	.0911 (2.57)**	.0517 (1.11)	.0474 (1.28)
7	–.0475	.113 (2.20)**	.0913 (2.57)**	.0497 (1.06)**	.0465 (1.25)
8	–.0448	.121 (2.44)**	.091 (2.64)***	.0488 (1.07)	.0429 (1.19)
9	–.0735	.135 (2.68)***	.0786 (2.31)**	.0515 (1.12)	.0413 (1.13)
10	.119	.106 (2.02)**	.108 (3.05)***	.0331 (.7)	.045 (1.19)

Contd...

S.N	S	P	E	D_E	EP
1	.000003 (1.977)*	.0011 (1.96)*	.0006 (.56)	–.013 (–1.21)	.0092 (2.74)***
2	.000004 (2.98)***	.0009 (1.57)	–.000 (–.001)	–.016 (–1.52)	.00795 (2.32)***
3	.000003 (2.10)**	.0012 (1.20)**	.0010 (.84)	–.006 (–.604)	.01097 (3.23)***
4	.000003 (2.02)**	.0011 (1.98)*	.0007 (.601)	–.0134 (–1.28)	.00883 (2.64)***
5	.000003 (2)**	.00114 (1.97)*	.0004 (.39)	–.0114 (–1.09)	.00921 (2.73)***
6	–	.0013 (2.21)**	.00056 (.475)	–.0001 (–.94	.0108 (3.27)***
7	.000003	–	.00081	–.0147	.00904

	(2.23)**		(.689)	(–1.39)	(2.64)***
8	.000003 (1.97)*	.00116 (2.01)**	–	-.0127 (–1.22)	.00905 (2.71)***
9	.000003 (1.83)*	.0012 (2.08)*	.00065 (.563)	–	.0097 (2.90)***
10	.000004 (2.62)***	.0011 (1.81)*	.00033 (.279)	–.0162 (–1.49)	–

Contd...

S.N	R^2	R^2	**F-value**	**Sig. Level**
1	.395	.328	5.878	.000
2	.349	.285	5.487	.000
3	.347	.283	5.444	.000
4	.387	.328	6.471	.000
5	.383	.323	6.373	.000
6	.366	.304	5.914	.000
7	.366	.305	5.926	.000
8	.393	.334	6.63	.000
9	.384	.324	6.394	.000
10	.389	.274	5.254	.000

Note: 1. T-Values are given in the parentheses; 2. *Significant at 10% level of significance; **Significant at 5% level of significance; *** Significant at 1% level of significance.

Table 3: Correlation Matrix (Type I_a equation)

	D_S	D_{MP}	D_{FA}	D_{BH}	S	P	E	D/E	EP
D_S	1								
D_{MP}	.299***	1							
D_{FA}	–.212*	–.205	1						
D_{BH}	–.384**	.038	–.203	1					
S	.402**	.227	–.102	–.103	1				
P	–.137	–.013	.062	.044	.057	1			
E	–.225*	–.003	.134	–.047	–.134	.103	1		
D/E	–.057	.227*	–.113	.175	.090	–.067	.000	1	
EP	.093	.281**	–.193	.053	.260*	.027	–.078	.046	1

Contd…

Note: *Correlation is significant at 5% level; **Correlation is significant at 1% level.

Table 4: Correlation Matrix (Type I_b equation)

	D_S	D_{MP}	D_{FA}	D_{BH}	S	P	E	D/E	EP
D_S	1								
D_{MP}	.299**	1							
D_{FA}	–.212*	–.205	1						
D_{BH}	–.384**	.038	–.203	1					
S	.402**	.227	–.102	–.103	1				
P	–.137	–.013	.062	.004	.057	1			
E	–.225*	–.003	.134	–.047	–.134	.103	1		
D/E	–.057	.227*	–.113	.175	.09	–.067	.000	1	
EP	.039	.215**	–.148	.054	.246*	.002	–.106	–.023	1

Contd…

Note: *Correlation is significant at 5% level; **Correlation is significant at 1% level.

Table 5: Testing for Superfluous Variables (Type II_c equation)

S. N	Constant	D_S	D_{FA}	D_{BH}	S
1	–.0856	.0844 (1.66)	.0543 (1.23)	.0316 (.871)	.000002 (1.24)
2	–.0574	–	.0373 (.855)	.0032 (.10)	.000002 (1.0)
3	–.0595	.070 (1.41)	–	.0203 (.578)	.000002 (1.318)
4	–.0701	.0636 (1.42)	.0445 (1.04)	–	.000002 (1.206)
5	–.0991	.0974 (1.95)*	.0578 (1.30)	.0296 (.813)	–
6	–.0811	.0701 (1.41)	.0564 (1.27)	.0292 (.802)	.000002 (1.33)
7	–.0756	.0781 (1.56)	.0557 (1.26)	.0281 (.784)	.000002 (1.23)
8	–.0935	.0863 (1.696)*	.0581 (1.31)	.0256 (.712)	.000002 (1.09)
9	.0888	.0766 (1.397)*	.0285 (.603)	.033 (.844)	000003 (2.0)**
10	–.0663	.0973 (1.88)*	.0476 (1.07)	.0432 (1.17)	.000002 (1.34)
11	–.732	.0882 (1.73)*	.0546 (1.23)	.031 (.851)	.000002 (1.19)
12	–.054	.145 (2.89)***	.046 (.977)	.0524 (1.39)	.000003 (2.06)**
13	–.0518	.0821 (1.60)	.0529 (1.18)	.0267 (.732)	.000002 (1.19)

Contd...

S. N	P	E	D/E	EP	DI_a
1	.00078 (1.36)	.0008 (.718)	–.0115 (–1.15)	.0182 (3.77)***	.117 (2.35)**
2	.00059 (1.08)	.0005 (.433)	–.012 (–1.19)	.0179 (3.66)***	.126 (2.52)**
3	.0008 (1.39)	.0009 (.770)	–.0124 (–1.244)	.0173 (3.61)***	.113 (2.27)**
4	.0008	.0007	–.0102	.0183	.123 (2.5)**

	(1.32)	(.608)	(–1.04)	(3.78)***	
5	.0008 (1.44)	.00076 (.679)	–.0098 (–.985)	.0196 (4.15)***	.121 (2.42)**
6	–	.00089 (.791)	–.0123 (–1.23)	.0182 (3.74)***	.123 (2.47)**
7	.0008 (1.40)	–	–.0114 (–1.14)	.0180 (3.74)***	.122 (2.47)**
8	.0008 (1.42)	.00078 (.698)	–	.018 (3.78)***	.110 (2.23)**
9	.0008 (1.23)	.0006 (.458)	–.0121 (–1.12)	–	.140 (2.62)**
10	.0009 (1.53)	.00113 (.991)	–.0087 (–.852)	.0196 (3.97)***	–
11	.0008 (1.42)	.0009 (.835)	–.0099 (–1.00)	.0192 (4.00)***	.089 (1.98)**
12	.001 (1.03)*	.0008 (.653)	–.0108 (–1.03)	.0189 (3.67)***	.051 (1.05)
13	.0007 (1.15)	.0008 (.695)	–.0083 (–.842)	.0184 (3.76)***	.0815 (1.80)*

Contd…

S. N	DI_c	DI_d	DI_e	R^2	R^2	F-value
1	.0621 (1.31)	.185 (3.26)***	.0682 (1.65)	.45	.365	5.31***
2	.0666 (1.39)	.220 (4.11)***	.0663 (1.59)	.43	.351	5.42***
3	.0624 (1.31)	.181 (3.18)***	.0672 (1.62)	.439	.361	5.62***
4	.0616 (1.3)	.194 (3.48)***	.0652 (1.59)	.444	.367	5.741***
5	.0598 (1.25)	.204 (3.71)***	.0 671 (1.62)	.439	.361	5.613***
6	.0656 (1.38)	.196 (3.45)***	.0612 (1.48)	.437	.358	5.567***
7	.0652 (1.38)	.185 (3.27)***	.0679 (1.65)	.446	.369	5.781***
8	.0557 (1.18)	.184 (3.23)***	.059 (1.45)	.44	.362	5.648***

Contd…

9	.090 (1.78)**	.193 (3.15)***	.0713 (1.6)	.349	.259	3.857***
10	.0153 (.344)	.131 (2.45)**	.0259 (.676)	.411	.328	5.002***
11	–	.159 (2.98)***	.0432 (1.17)	.438	.359	5.587***
12	.0071 (.151)	–	.0192 (.47)	.375	.287	4.301***
13	.026 (.60)	.151 (2.83)***	–	.430	.351	5.427***

Note: 1. T-Values are given in the parentheses; 2. *Significant at 10% level of significance; **Significant at 5% level of significance; *** Significant at 1% level of significance.

Table 6: Testing for Superfluous Variables (Type II_d equation)

S.N	Constant	D_S	D_{FA}	D_{BH}	S
1	–.105	.112 (2.14)**	.0516 (1.134)	.0342 (.915)	.000002 (1.27)
2	–.0606	–	.0286 (.633)	.0032 (–.08)	.000003 (1.73)*
3	–. 0792	.098 (1.92)*	–	.0234 (.647)	.000002 (1.34)
4	–.0888	.0896 (1.94)*	.041 (.933)	–	.000002 (1.23)
5	–.127	.126 (2.46)**	.0552 (1.21)	.032 (.86)	–
6	–.0956	.0952 (1.84)*	.0537 (1.17)	.0313 (.832)	.000002 (1.38)
7	–.0927	.106 (2.06)**	.0528 (1.17)	.031 (.833)	.000002 (1.26)
8	–.116	.114 (2.18)**	.055 (1.22)	.029 (.80)	.000002 (1.14)
9	.0981	.098 (1.78)*	.0326 (.684)	.0355 (.901)	.000003 (2.07)**
10	–.0835	.127 (2.36)**	.043 (.922)	.0471 (1.23)	.000002 (1.38)
11	–.0937	.118 (2.23)**	.0518 (1.13)	.0333 (.882)	.000002 (1.17)
12	–.0649	.170 (3.30)***	.0421 (.882)	.0545 (1.41)	.000003 (2.09)**
13	–.0634	.109 (2.06)**	.0495 (1.08)	.0289 (.766)	.000002 (1.23)

Contd…

S.N	P	E	D_E	EP	DI_a
1	.00094 (1.58)	.0008 (.69)	–.0093 (–.901)	.0106 (3.17)***	.130 (2.54)**
2	.00068 (1.14)	.00036 (.31)	–.0101 (–.96)	.010 (2.94)***	.142 (2.73)***
3	.0010 (1.612)	.00085 (.733)	–.0103 (–.997)	.0101 (3.05)***	.126 (2.46)**
4	.00009	.0007	–.0079	.0106	.136 (2.69)***

	(1.54)	(.574)	(–.778)	(3.19)***	
5	.00099 (1.68)*	.00078 (.669)	–.00723 (–.708)	.0117 (3.62)***	.134 (2.61)**
6	–	.00089 (.764)	–.0103 (–.993)	.0104 (3.07)***	.137 (2.67)***
7	.00096 (1.62)	–	–.0092 (–.898)	.0104 (3.13)***	.134 (2.66)***
8	.00097 (1.64)	.00079 (.684)	–	.0109 (3.27)***	.124 (2.45)**
9	.00084 (1.36)	.0004 (.333)	–.0121 -(1.11)	–	.144 (2.68)**
10	.0011 (1.76)*	.0012 (.99)	–.006 (–.57)	.0114 (3.3)***	–
11	.00099 (1.67)*	.00099 (.853)	–.0067 (–.675)	.0116 (3.47)***	.0928 (1.97)*
12	.00117 (1.9)*	.00175 (.619)	–.0087 (–.806)	.0104 (2.95)***	.0664 (1.35)
13	.00081 (1.36)	.00077 (.657)	–.0059 (–.573)	.0104 (3.09)***	.0912 (1.95)*

Contd...

S.N	DI_c	DI_d	DI_e	R^2	R^2	F-value
1	.086 (1.75)*	.180 (3.07)***	.0746 (1.75)*	.431	.344	4.927***
2	.092 (1.84)*	.225 (4.03)***	.072 (1.65)	.398	.314	4.744***
3	.0857 (1.75)*	.175 (3.00)***	.0733 (1.72)*	.422	.341	5.238***
4	.0849 (1.74)*	.189 (3.29)***	.0714 (1.68)*	.425	.345	5.308***
5	.0825 (1.68)*	.20 (3.54)***	.074 (1.73)*	.419	.339	5.189***
6	.0902 (1.83)*	192 (3.28)***	.0661 (1.55)	.413	.331	5.051***
7	.089 (1.83)*	.179 (3.07)***	.0741 (1.75)*	.428	.348	5.367***
8	.0799 (1.65)	.179 (3.06)***	.0673 (1.61)	.425	.345	5.313***

Contd...

9	.111 (2.18)**	.176 (2.84)***	.0706 (1.57)	.358	.268	3.999***
10	.0338 (.734)	.120 (2.163)**	.0279 (.703)	.384	.298	4.479***
11	–	.145 (2.6)**	.0408 (1.062)	.409	.327	4.969***
12	.0336 (.695)	–	.027 (.648)	.362	.274	4.081***
13	.0466 (1.05)	.143 (2.58)**	–	.409	.326	4.966***

Note: 1. T-Values are given in the parentheses; 2. *Significant at 10% level of significance; **Significant at 5% level of significance; *** Significant at 1% level of significance.

Table 7: Correlation Matrix (Type II_c equation)

	D_S	D_{FA}	D_{BH}	S	P	E
D_S	1					
D_{FA}	–.212*	1				
D_{BH}	–.384**	–.203	1			
S	.402**	–.102	–.103	1		
P	–.137	.062	.044	.057	1	
E	–.225*	.134	–.047	–.134	.103	1
D_E	–.057	–.113	.175	.09	–.067	.000
EP	.093	–.193	.053	.26*	.027	–.078
DI_a	–.059	–.097	.174	.025	.105	.113
DI_c	–.072	.057	–.037	–.93	.072	.106
DI_d	–.476**	–.182	–.015	.443**	.079	–.185
DI_e	–.131	.067	–.086	–.156	–.222*	–.029

Contd...

	D/E	EP	DI_a	DI_c	DId	DI_e
D_S						
D_{FA}						
D_{BH}						
S						
P						
E						
D_E	1					
EP	.046	1				
DI_a	.084	.110	1			
DI_c	.010	.098	–.189	1		
DI_d	–.035	.110	–.182	–.189	1	
DI_e	.128	–.120	–.270**	– .281**	– .270**	1

Note: *Correlation is significant at 5% level; **Correlation is significant at 1% level.

Table 8: Correlation Matrix (Type II_d equation)

	D_S	D_{FA}	D_{BH}	S	P	E
D_S	1					
D_{FA}	–.212*	1				
D_{BH}	–.384**	–.203	1			
S	.402**	–.102	–.103	1		
P	–.137	.062	.044	.057	1	
E	–.225	.134	–.047	–.134	.103	1
D/E	–.057	–.113	.175	.090	–.067	.000
EP	.039	–.148	.054	.246*	.002	–.106
DI_a	–.059	–.097	.174	.025	.105	.113
DI_c	–.072	.057	–.037	–.093	.072	.106
DI_d	–.476**	–.182	–.015	.443**	.079	–.185
DI_e	–.131	.067	–.086	–.156	–.222*	–.029

Contd…

	D/E	EP	DI_a	DI_c	DI_d	DI_e
D_S						
D_{FA}						
D_{BH}						
S						
P						
E						
D/E	1					
EP	.023	1				
DI_a	.084	.086	1			
DI_c	.010	.139	–.189	1		
DI_d	–.035	.048	–.182	–.189	1	
DI_e	.128	–.145	–.270**	–.281**	–.270**	1

Note: *Correlation is significant at 5% level; **Correlation is significant at 1% level.

Bibliography

Bibliography

Books

Belkaoui A., *Accounting Theory*, Harcourt Brace, Jovanovich, (1981).

Bennett M., Rikhardsson P.M. and Schaltegger S., *Environmental Management Accounting – Purpose and Progress*, Kluwer Academic Publishers, The Netherlands, (2003).

Boyd H.W., Westffall R. and Stasch S.F., *'Marketing Research – Text and Cases'*, A.I.T.B.S. Publishers and Distributors, Delhi, 7th ed., (2002),

Cascio J., *The ISO 14000 Handbook*, CEEM Information Services, (1996).

Dierker M. and Bauer R.A., *Corporate Social Accounting*, Praeger Publishers, New York, U.S.A., (1973).

Estes P.W., *Accounting and Society*, Melville Publishing Company, Los Angeles, California, (1973).

Estes R.W., *Corporate Social Accounting*, John Wiley and sons, New York (1996).

Gray R. and Bebbington T., *Accounting for the Environment*, Sage Publications, London, 2nd edition, (2001).

Gray R.H., Bebbington K.J., and Walters D., *Accounting for the Environment*, Paul Chapman Publishers, London, (1993).

Gray R.H., Owen D.L. and Maunders K.T., *Corporate Social Reporting: Accounting and Accountability*, Prentice Hall, Hend Hempstead, (1987).

Gujrati Damodar N., *Basic Econometrics*, McGraw Hill Book Company, New York, (2003).

Hansen D.R. and Mowen M.M., *Cost Management Accounting and Control*, South Western College Publishing, Cincinnati, USA, 3rd edition, (2000).

Kmenta J., *Elements of Econometrics*, MacMillan Publishing Company, (1971).

Kumar A. and Goel P.K., *Industry, Environment and Pollution*, ABD Publishers, Jaipur, (2000), pp. 390-395.

Lal J., *Accounting Theory*, Himalaya Publishing House, Mumbai, 2nd ed., (2001).

Nobes C. and Parker, R. (eds.), *Comparative International Accounting*, Prentice Hall, Hemel Hempstead, UK, 3rd ed.,

(1991).

Owen D., *Green Reporting- Accountancy and the Challenges of the Nineties,* Chapman and Hall, London, UK, (1992),

Porwal L.S., *Accounting Theory*, Tata McGraw Hill Publishing Company Ltd., New Delhi, 3rd ed., (2001).

Pramanik A.K., *Environmental Accounting and Reporting*, Deep and Deep Publications Pvt. Ltd., New Delhi, (2003).

Rao P. and Miller R.L., *Applied Econometrics*, Prentice Hall of India Pvt. Ltd., New Delhi, (1972).

Rao P.S.B. and Rao, P.M., *Environmental Management and Audit*, Deep and Deep Publications Pvt. Ltd., New Delhi, (2000).

Rathore S., *International Accounting*, Prentice Hall of India Pvt. Ltd., New Delhi, (1996).

Saudagaran S.M., *International Accounting: A User Perspective*, Thomson Asia Pvt. Ltd., Singapore, (2001).

Truilt Thomas and others, *Environmental Audit Hand Book Basic Principles of Environmental Auditing*, Resource Planning Corporation, Washington, D.C., 2nd edition, (1999).

Walton P., Axel H. and Raffourmer B., *International Accounting*, International Thomson Business Press, (1998).

Articles and Reports

Abu-Ghazaleh T., "Financial Services and the Environment: The Accounting Profession", *UNEP Industry and Environment*, (Jan-March, 1999), pp. 32-34.

ACCA and UNEP 2002, '*Industry as a Partner for Sustainable Development Accounting*', A report prepared by Association of Chartered Certified Accountants (ACCA), developed through a Multi-stakeholder process facilitated by UNEP, UK, http://www.accaglobal.com, Accessed on April 15, 2007.

ACCA, "Mandatory Environmental Accounting in Malaysia", Source: www.acca.co. uk/pdfs/environment/other/TECH-ERM-001-5.pdf

Accounting Principles Board (APB), Statement No.4, *Basic Concepts and Accounting Principles Underlying Financial Statements of Business Enterprise*, AICPA, (1970), Para 40.

Al-Tuwaijri S.A., Christensen T.E. and Hughes II K.E., "The Relations Among Environmental Disclosure, Environmental Performance, and Economic Performance: A Simultaneous Equations Approach", *Accounting, Organizations and Society*,

29, (2004), pp. 447-471.

Ansari A.A., "Environmental Accounting and Reporting", *The Indian Journal of Commerce*, Vol. 54, No. 3, (July-Sept. 2001), pp. 84-96.

Bebbington J., "Compulsory Environmental Reporting in Denmark: An Evaluation", *Social and Environmental Accounting*, Vol. 19(2), (1999), pp. 2-4.

Bebbington J., "Corporate Environmental Reporting in Denmark: An Evaluation", *Social and Environmental Accounting*, Vol. 19(2), (1999), pp. 2-4.

Bebbington J., Gray R., Thomson I., and Walters, D., "Accountant's Attitudes and Environmentally Sensitive Accounting", *Accounting and Business Research*, Vol. 24, No.94, (Spring 1994), pp. 109-120.

Bebbington, J. and Larrinaga, C., "Carbon Trading: Accounting and Reporting Issues", *European Accounting Review*, Vol. 17, Issue 4, (Dec.2008), pp.697-717.

Belal A.R., "Corporate Social Reporting in Bangladesh", *Social and Environmental Accounting*, Vol.19 (1), (1999), pp.8-12.

Blacconiera W.G. and Patten D.M., "Environmental Disclosures, Regulatory Costs, and Changes in Firm Value", *Journal of Accounting and Economics*, 18(3), (1994), pp. 357-377.

Borghini S., 'Environmental Information in Annual Reports', *Environmental Communication Observatory*, No. 2, (Sep. 1998), Milane.

BT-TERI Survey, "India's Greenest Companies", *Business Today*, (May 6, 2001), pp. 64-74.

Buhr N. and Freedman M., "A Comparison of Mandated and Voluntary Environmental Disclosure", Source: http://panoptic.csustan.edu/cpa 96/txt/buhr.txt,198

Canadian Institute of Chartered Accountants, "*Environmental Costs and Liabilities: Accounting and Financial Reporting Issues*", CICA Toronto, (1993), p. 129.

Canadian Institute of Chartered Accountants, "*Reporting on Environmental Performance*", CICA, Toronto, (1994).

Cerin P., "Characteristics of Environmental Reporters on the OM Stockholm Exchange", *Business Strategy and the Environment*, 11, (2002), pp. 298-311.

Certified Management Accountant (CMA), "Management Accounting Guideline 40 – Tools and Techniques of

Environmental Accounting for Business Decisions", *CMA*, Canada.

Charumathi B., "Added Accountability of the Corporate Sector for Environmental Concerns", *Journal of Accounting and Finance*, Vol. 12, No. 2, (Sept. 1998).

Clarks J. and Gibson M., "Use of Corporate Social Disclosures in the Management of Reputation and Legitimacy: A Cross Sectoral Analysis of UK Top 100 Companies", *Business Ethics – A European Review*, Vol. 18, No. 1, (Jan. 1999), pp. 5-13.

Darryl R., Beth, Beloff and Darryl D., "Case Studies on Corporate Environmental Accounting", Source: http://Green Ledgers in Corporate Environmental Accounting, htm.

David W., "Legal Consideration in Voluntary Corporate Environmental Reporting", *ELR – News and Analysis*, Environmental Law Institute, Washington, (2000).

Deegan C. and Gordon B., "A Study of the Environmental Disclosure Practices of Australian Corporations", *Accounting and Business Research*, Vol. 26, No. 3, (1996), pp. 187-189.

Deegan C. and Rankin M., "Do Australian Companies Report Environmental News Objectively?", *Accounting, Auditing & Accountability Journal*, Vol. 9, No. 2, (1996), pp. 50-67.

Depoers F., "An Empirical Investigation of The Environmental Disclosure Strategy of French Listed Companies", a paper presented at Euro conference project, Financial Reporting and Recognition Practices in Europe (FIRRE), (May 20-23,2001), Altavilla Milicia, (Sicily, Italy).

DTTI, IISD, & Sustain Ability, "Coming Clean – Corporate Environmental Reporting, Opening up for Sustainable Development", (1993), DTTI: London.

Emtairah T., "Corporate Environmental Reporting – Review of Policy Action in Europe", *IIIEE Report*, Lund (Feb. 2002).

Environmental Protection Agency (EPA), '*An Introduction to Environmental Accounting as a Business Management Tool: Key Concept sand Terms*', US: EPA primer, Washington DC: EPA, (1995a)

Environmental Protection Agency's (EPA) document "Environmental Accounting Case Studies", Green Accounting , AT&T, (Sept. 1995).

Eresi K., "Information Disclosers in Annual Reports, *The Chartered Accountant*, (Jan. 1996), pp. 45-48.

Ernst and Ernst, "Accounting for the Corporate Social Performance: Measuring Costs of Social Action", A Study Conducted for N.A.A. (U.S.A.), *Management Accounting*, (Sept. 1975), pp. 60-64.

EU Recommendation on Measurement and Disclosure on the Recognition of Environmental News in the Annual Accounts and Annual Reports of Companies, May 30, 2001, Source: http://europa.eu.ent/ comm/internal-market

FASB, *Conceptual Framework for Financial Accounting and Reporting*, (1976)

Federation Des Experts Comptables Europeens (FEE), *'Discussion Paper Towards a Generally Accepted Framework for Environmental Reporting'*, (1999), Source: http://www.fee:brussels.fee.be/publications/main.htm.

Ford R.L., "Accountants and the Environment", *Accountancy*, (June 1995), pp. 128-129.

Frankiss W., "Seen to be Green", *CA Environment*, (May 2001), pp. 40-41.

Freedman M. and Jaggi B., "Pollution Disclosure, Pollution Performance and Economic Performance", *Omega*, 10, (1982), pp. 167-176.

Frost G.R., "An Investigation of the Introduction of Mandatory Environmental Reporting in Australia", 3rd Asia Pacific Interdisciplinary Research in Accounting (APIRA) Conference, (15-17 July 2001), Adelaide.

Frost G.R., "Environmental Reporting in Australian Extractive Industry", (1998), Source: http://www.bham.ac.uk/EAA/eaa98/abstracts/sea/frost.htm.

Gamble G.O., Hsu K., Kite D. and Radtke R., "Environmental Disclosures in Annual Reports and 10 Ks: An Examination", *Accounting Horizons*, 9(3),(1995), pp.34-54.

Global Reporting Initiative (GRI), *Sustainability Reporting Guidelines*, 1991, GRI: Boston.

Gonzalez C.L., "Steps Towards Mandatory Environmental Accounts in Spain", *Social and Environmental Accounting*, Vol. 19, No. 1, (1999), p. 11.

Gray R., Jarad M., Power D. and Sindair D., "Social and Environmental Disclosure and Corporate Characteristics: A Research Note and Extension", *Journal of Business Finance and Accounting*, 28(3)&(4), (April/May 2001), pp. 327-356.

Gray R., Kouhy R. and Lavers S., "Corporate Social and Environmental Reporting: A Review of the Literature and a Longitudinal Study of UK Disclosure", *Accounting, Auditing & Accountability Journal*, Vol. 8, No. 2, (1995a), pp. 47-77.

Gray R., Kouhy R. and Lavers, S., "Methodological Themes: Constructing a Research Database of Social and Environmental Reporting by UK Companies", *Accounting, Auditing & Accountability Journal*, Vol. 8, No. 2, (1995b), pp. 78-101.

Guidelines issued on Global Reporting Initiative (GRI), Source: www.globalreporting.org.

Guthrie J. and Parker L., "Corporate Social Disclosure Practice: A Comparative International Analysis", *Advances in Public Interest Accounting*, Vol. 3, (1990), pp. 159-76.

Guthrie J. and Parker L., "Corporate Social Reporting: A Rebuttal of Legitimacy Theory", *Accounting and Business Research*, Vol. 9, Issue 76, (1989), pp. 343-356.

IASC, "A Framework for the Preparation and Presentation of Financial Statements", *The Chartered Accountant*, (Aug. 1989), pp. 124-133.

Imam A., "Environmental Reporting in Bangladesh", *Social and Environmental Accounting,* Vol. 19(2), (1999), pp.12-14.

Ince D., "Determinants of Social and Environmental Disclosures in the UK Companies in Environmental Policy Statements", Source: http://les.man.ac.uk/IPA97/ papers/ince119.pdf

Ingram R. and Frazier K., "Environmental Performances and Corporate Disclosure", *Journal of Accounting Research*, 18, Autumn, (1980), pp. 612-622

Jaffar R., Iskandar T.M. and Muhamad M., "An Investigation of Environmental Disclosure in Malaysia", Source: www.aaanz.org/web2002/accepted0.020-papers/jafar.pdf, as last accessed on 30-1-05.

Jaggi B. and Freedman M., "An Examination of the Impact of Pollution Performance on Economic and Market Performance of Pulp and Paper Firms", *Journal of Business Finance and Accounting*, 19(5), (1992), pp. 697-713.

Johnson L.T., "Research on Environmental Reporting" *Accounting Horizon*, 5(3), (Sept. 1993), pp. 118-123.

Jones K., '*Study on Environmental Reporting by Companies*', Commission of European Communities, DG Environment, Centre for Environmental Informatics, University of Sunderland,

(March 2000).

Kini N., "Green Accounting", *The Chartered Accountant*, (Nov. 1998), pp. 36-38.

Kolk A., Walhain S. and Wateringen S.V., "Environmental Reporting by the Fortune Global 250: Exploring the Influence of Nationality and Sector", *Business Strategy and the Environment*, 10, (2001), pp. 15-28.

KPMG, '*International Survey on Corporate Responsibility Reporting*, 2005', Source: http://www.kpmg.nl/Docs/Corporate_site/Publicates/International_Survey_Corporate_Responsibility_2005.pdf.

KPMG, '*KPMG International Survey of Corporate Sustainability Reporting 2002*', (2002), Source: http://www.wimm.nl/publicates/KPMG2002.pdf

KPMG, '*KPMG International Survey of Environmental Reporting, 1996*', Source: www.wimm.hl/ publication/kpmg 1999.pdf

Larrinaga C., "An Evaluation of Mandatory Environmental Accounts in Spain", Source: visar.csustan.edu/papers/larrinage 91.pdf

Lober Douglas, J., "The 100+ Corporate Environmental Report Study: A Survey of an Evolving Environment Tool", *Business Strategy and the Environmental*, (1997), pp. 57-59.

Lodhia, S.K. 'Accountants' Responses to the Environmental Agenda in a Developing Nation: An Initial Study on Fiji' (2000), Source: http://ecocomm.anu.edu.au/ bim/staff/lodhia/lodhia.html

Mahmud H., Tan L.M. and Adams M., "Voluntary Disclosure in an Emerging Capital Market: Some Empirical Evidence from Companies Listed on Kuala Lumpur Stock Exchange," *The International Journal of Accounting*, 29, (1994), pp. 334-351.

Maji M.M., "Environmental Accounting and Reporting – An Emerging Issue", *The Indian Journal of Commerce*, Vol. 53, No. 1&2, (Jan-June 2000), pp. 107-114.

Martin and others, "*An FEE Approach to Environmental Accounting*", Source: http://www.academy.com

Moneva J.M. and Llena F., "Environmental Disclosures in the Annual Reports of Large Companies in Spain", *The European Accounting Review*, 9:1, (2000), pp. 7-29.

Nagle G., "Business Environmental Cost Accounting Survey," Global Environmental Management Initiative, 1994 Conference proceedings, Arlington,(March 16-17,1994), p. 243.

Nandan R.K., "Accounting and the Environmental Issues of Today:

A Shift Towards Green Thinking in Accounting", *The Fiji Accountant*, (1993), pp. 4-10.

National Association of Accountants' (NAA) Committee Report on Accounting for Corporate Social Performance, (1976).

Niskala M. and Pretes M., "Environmental Reporting in Finland", *Accounting, Organizations and Society*, 20(6), (1995), pp. 457-466.

Nitkin D. and David P., (1993), "Corporate Sustainable Development Reporting in Canada", Working Paper No. 17, National Round Table on Environment and Economy (NRTEE) Ottawa: NRTEE.

O'Dwyer B. and Gray R.H., "Corporate Social Reporting in the Republic of Ireland: A Longitudinal Study", *The Irish Accounting Review*, 5(2), (1998), pp. 1-34.

O'Dwyer B., "Managerial Perceptions of Corporate Social Disclosure: An Irish Story", *Accounting, Auditing & Accountability Journal*, 15(3), (2002), pp. 403-436.

O'Dwyer B., "The Ponderous Evolution of Corporate Environmental Reporting in Ireland: Recent Evidence from Publicly Listed Companies", *Corporate Social Responsibility and Environmental Management*, 10, (2003), pp. 91-100.

Owen D.L., "The Need for Environmental Accounting Standards", *Accounting Forum*, Vol. 17, No. 4, (March 1994), pp. 31-46.

Pahuja S. and Bansal S.C., "*Environmental Accounting in an Emerging Economy: A Study of Large listed Companies in India*", a Paper presented at 29th Annual congress of European Accounting Association (EAA), held at Dublin, Ireland, (March 22-24. 2006).

Pahuja S., "Accountants' Response to Environmental Accounting and Reporting Challenges in an Emerging Economy: A Study of India", in Sahey, Stough et al., *Green Business*, New Delhi: Allied Publishers Pvt. Ltd., (2006), 281-310.

Pahuja S., "Corporate Environmental Reporting Expectation Gap: Evidence from India", *The ICFAI Journal of Accounting Research*, VI (2), (April 2007), pp.21-42.

Palash Goswami, "Accounting and Reporting of Carbon Credits and Their Trading", *The Accounting World*, (Oct. 2008).

Patten D.M., "Intra-industry Environmental Disclosures in Response to the Alaskan Oil Spill: A Note on Legitimacy Theory", *Accounting, Organizations and Society*, Vol. 15, No. 5, (1992), pp. 471-475.

Pensions and Investment Research Consultants (PIRC), 'Environmental Reporting 1996: The PIRC Survey of FTSE 350', PIRC, (1996)

Piet J., "Guarding the Globe", *Accountancy*, (2000), pp. 62-63.

Pinazza A., "Green Accounting", *CA Charter*, (2000), pp. 60-63.

Porwal L.S. and Sharma N., "Social Responsibility Disclosure by Indian Companies", *The Chartered Accountants*, (Feb 1989), pp. 630-635.

Ratnatunga, J., "An Inconvenient Truth about Accounting", *Journal of Applied Management Accounting Research, 5*(1), (2007).

Rezaee Z., "Help Keep the World Green", *Journal of Accountancy*, (Nov. 2000), pp. 57-79.

Rikardsson P.M., "The Measurement and Reporting of Corporate Environmental Information", A Paper prepared for the Nordic Network Conference, 'Business and the Environment, Sandvika, 7-12 Dec. 1994'.

Roberts C.B., "Environmental Disclosures: A Note on Reporting Practices in Mainland Europe", *Accounting, Auditing & Accountability Journal*, Vol. 4, No. 3, (1991), pp. 62-71.

Scott P., "ISO Backs Environmental Reporting", *Environmental Finance*, (Nov. 2001), pp. 20-21.

Scott P., "Reporting All Over the World", *Environmental Finance*, (Dec. 2000 – Jan 2001), pp. 36-37.

Simster M., "How Green is my Tally?", *Charter*, (Aug. 2000), pp. 53-55.

Singh D.R. and Ahuja J.M. – "Corporate Social Reporting" *International Journal of Accounting*, (March 1983), pp. 151-169

Solomon A. and Lewis L., "Incentives and Disincentives for Corporate Environmental Disclosure", *Business Strategy and the Environment*, 11, (2002), pp. 154-169.

Stanko Brian B. and Zeller T.L., "Environmental Liability in Financial Reporting", *Business and Economic Review*, Vol. 41, Issue 2, (Jan-March 1995), pp. 19-23.

Stanwick P.A. and Stanwick S.D., "CEO Compensation: Does it pay to be Green?", *Business Strategy and the Environment*, Vol. 10, (2001), pp. 176-182.

Stanwick S.D. and Stanwick P.A., "The Relationship Between Environmental Disclosures and Financial Performance: An Empirical Study of US Firms", *Eco-Management and Auditing*, 7, 4, Dec. (2000) pp. 155-164.

Stratos, '*A Snapshot of Sustainability Reporting in Canada*', Canada, (2001), Source: www.stratos-sts.com/sts-files/stratos.chapter.2.pdf

Stray S. and Ballantine J., "A Sectoral Comparison of Corporate Environmental Reporting and Disclosures", *Eco-Management and Auditing*, 7, (Dec. 2000), pp. 165-177.

Surma J.P. and Vondra A.A., "Accounting for Environmental Costs: A Hazardous Subject," *Journal of Accountancy*, (March 1992), pp. 51-55.

The Environment Agency, *Standards and Guidelines*, 2004, Source: http:/ environmentagency.gov.uk/business

Thirumoorthy P., "Taking Professional Care of the Nature: The Environmental Audit", *The Chartered Accountant*, (Aug. 2002), pp. 148-155.

Thomas P.B. and Kenney S.Y., 'Environmental Reporting: A Comparison of Annual Report Disclosures and Popular Financial Press Commentary', (1997), Source: http://les.man.ac.uk/ipa97/ papers/thomas61.html

Thomas P.B., 'An Exploratory Study of the Extent of Environmental Disclosures by Multinational Corporations', (1996), Source: http://www3.bus.osaka-cu.ac.jp/cpa96/papers.htm/thomas.htm

Tilt C.A., "The Content and Disclosure of Australian Corporate Environmental Policies", Source: www.ssh.flinders.edu.au/commerce/researchpapers/aa.4.htm

Trotman K.T., "Social Responsibility Disclosures by Australian Companies", *The Chartered Accountants in Australia*, (March 1979), pp. 24-29.

UN-ICAI Seminar Report, "*Environmental Accounting and Reporting*", conducted at New Delhi, (August 2-4, 1999).

United Nations, Environmental Financial Accounting Guidelines, Reports Prepared for the UNCTAD Intergovernmental Working Groups of Experts on International Standard of Accounting and Reporting (ISAR), New York, UN, (1997).

United Nations Environment Programme (UNEP), '*Environmental Auditing*', Technical Report Series No. 2, UNEP, Industry and Environment office, (1990).

United Nations, '*Environmental Accounting : Current Issues, Abstracts and Bibliography*', United Nations, New York , (1992a).

United Nations, '*Environmental Disclosures: International Survey of*

Corporate Reporting Practices', Report of the UNCTAD Secretariat, (E/C.A.s10/A(3)1994/4), New York, (12 Jan.1994).

United Nations, "*Accounting And Financial Reporting For Environmental Costs And Liabilities*", A Manual prepared for the United Nations Conference on Trade and Development (UNCTAD) under the guidance of an International review group and technical support of UNCTAD/ISAR: Intergovernmental Working Group of Experts on International Standards of Accounting and Reporting (ISAR), CICA: Canadian Institute of Chartered Accountants, ACCA: *Association of Chartered Certified Accountant, UK, (*2000).

Vasal V., "Accounting for Sustainable Economic Development Issues and Evidences, *Indian Journal of Accounting*, Vol. XXX1, (June, 1998).

Wagner M., Phu N.V., Azomahou T. and Wehrmeyer W., "The Relationship between the Environmental and Economic Performance of Firms: An Empirical Analysis of the European Paper Industry", *Corporate Social Responsibility and Environmental Management*, 9, (Sept. 2002), pp. 133-146.

Wheeler D., "The End of the Corporate Environmental Report", *Business Strategy and the Environment*, (2001).

White A.L., "Sustainability and the Accountable Corporation", *Environment*, (Oct. 1999), pp. 30-41.

Wilmhurst T. and Frost R., "Environmental Accounting – a growing concern", *Australian CPA*, (May 1998), Source: http://www.environment.gov.au/epg/ environet/eecp/tools4.html

Wilmhurst T., '*A Review of Environmental Accounting Issues for Australian Accountants*', cited in a conference paper, prepared for the Australian Chamber of Manufacturers, Environmental Accounting and Reporting Conference, ASCPA, (19th June, 1997), See http://www.ea.gov.au.

Websites

- http:/www.environment agency.gov.uk/business
- www.sustainability.com.
- http:///www.ea.gov.au.
- www.globalreporting.org.
- http://www.enviroreporting.com
- Environmental accounting.htm

Index

Index